J.J. Pan and Partners 潘冀联合建筑师事务所

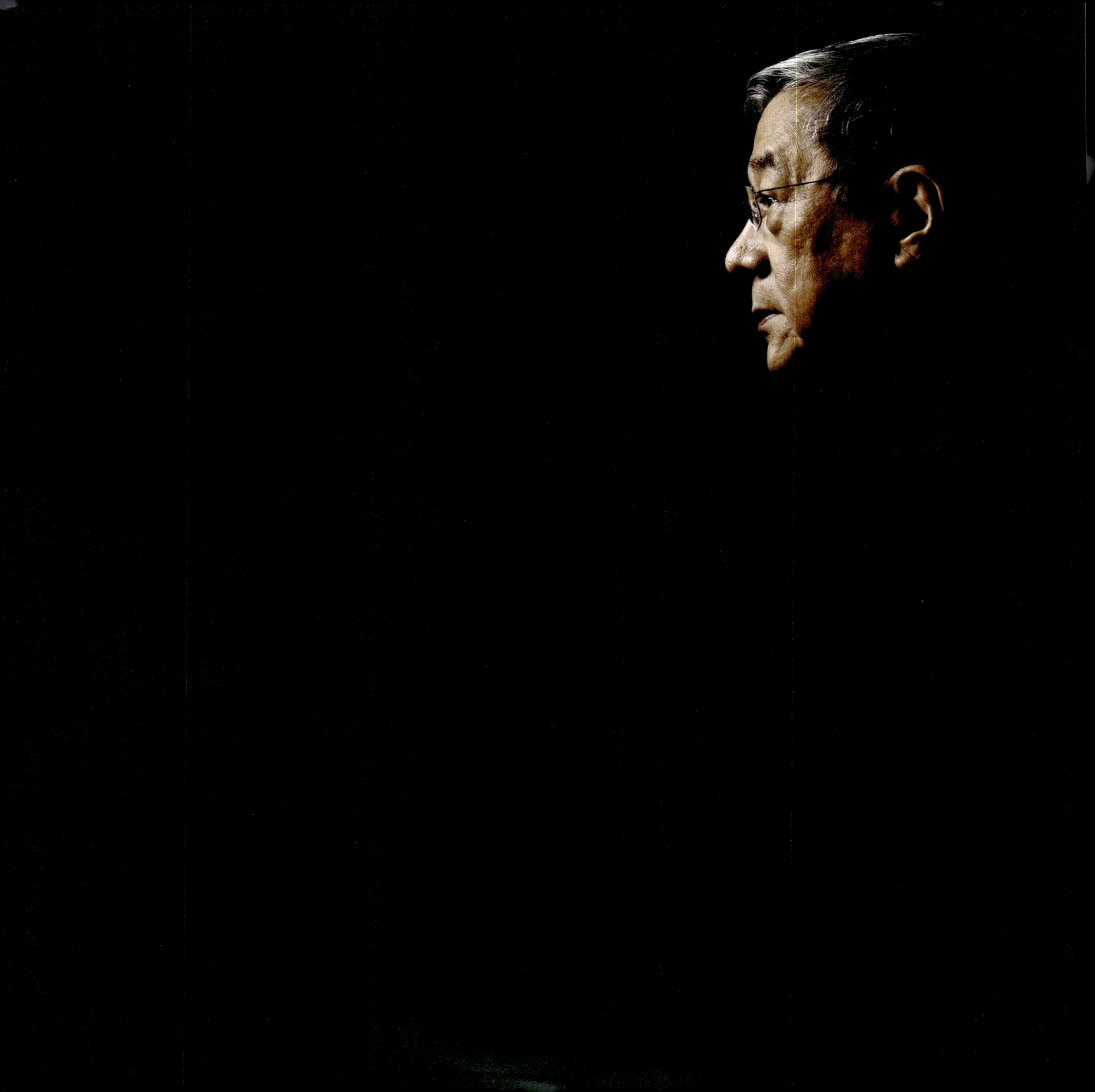

Cultivating Compassion
人本建筑

Humanistic Architecture as Practiced by JJP Architects and Planners
潘冀联合建筑师事务所的建筑实践

by JJP 潘冀联合建筑师事务所 / 编著

TONGJI UNIVERSITY PRESS 同济大学出版社

图书在版编目（CIP）数据

人本建筑：潘冀联合建筑师事务所的建筑实践 / 潘冀联合建筑师事务所编著 .– 上海：同济大学出版社，2018.10

ISBN 978-7-5608-7904-8

Ⅰ.①人… Ⅱ.①潘… Ⅲ.①建筑设计－作品集－中国－现代 Ⅳ.① TU206

中国版本图书馆 CIP 数据核字 (2018) 第 128184 号

人本建筑

潘冀联合建筑师事务所的建筑实践

潘冀联合建筑师事务所　编著

出 版 人：华春荣
策　　划：潘冀、 秦蕾 / 群岛工作室
统　　筹：苏重威
项目负责：王钟敏
JJP 团队：潘冀、苏重威、王钟敏、李大威、程露华、蔡君阳、凌琳、王维康、萧钧豪、翁思严、廖志坚、冷雪霜、黄文萦
平面设计：typo_d、蔡君阳、王维康、VV_design
英文翻译：李大威、Simon Frank、王钟敏、刘宛妮、金梦、梁廷达
责任编辑：晁艳

版　　次：2018 年 11 月第 1 版
印　　次：2018 年 11 月第 1 次印刷
印　　刷：上海雅昌艺术印刷有限公司
开　　本：787mm×1092mm 1/12
印　　张：22 1/3
字　　数：563 000
书　　号：ISBN 978-7-5608-7904-8
定　　价：268.00 元
出版发行：同济大学出版社
地　　址：上海市杨浦区四平路 1239 号
邮政编码：200092
网　　址：http://www.tongjipress.com.cn
经　　销：全国各地新华书店
光明城联系方式：info@luminocity.cn

Cultivating Compassion

Humanistic Architecture as Practiced by JJP Architects and Planners

ISBN 978-7-5608-7904-8
Initiator: Joshua Jih Pan, Qin Lei / Studio Archipelago
Coordinator: Chungwei Su
Manager: Chungmin Wang
JJP Team: Joshua Jih Pan, Chungwei Su, Chungmin Wang, David Lee, Lu Cheng, Daniel Y. Tsai, Lin Ling, Wei-Kang Wang, Chun-Hao Hsiao, Shih Yen Weng, Frank C.C. Liao, Syue-Shuang Leng, Vianne W.Y. Huang
Design: typo_d, Daniel Y. Tsai, Wei-Kang Wang, VV_design
Translator: David Lee, Simon Frank, Chungmin Wang, Debbie Liu, Meng Jin, Tingda Liang
Editor: Chao Yan
Published in November 2018, by Tongji University Press, 1239, Siping Road, Shanghai, China, 200092.
www.tongjipress.com.cn

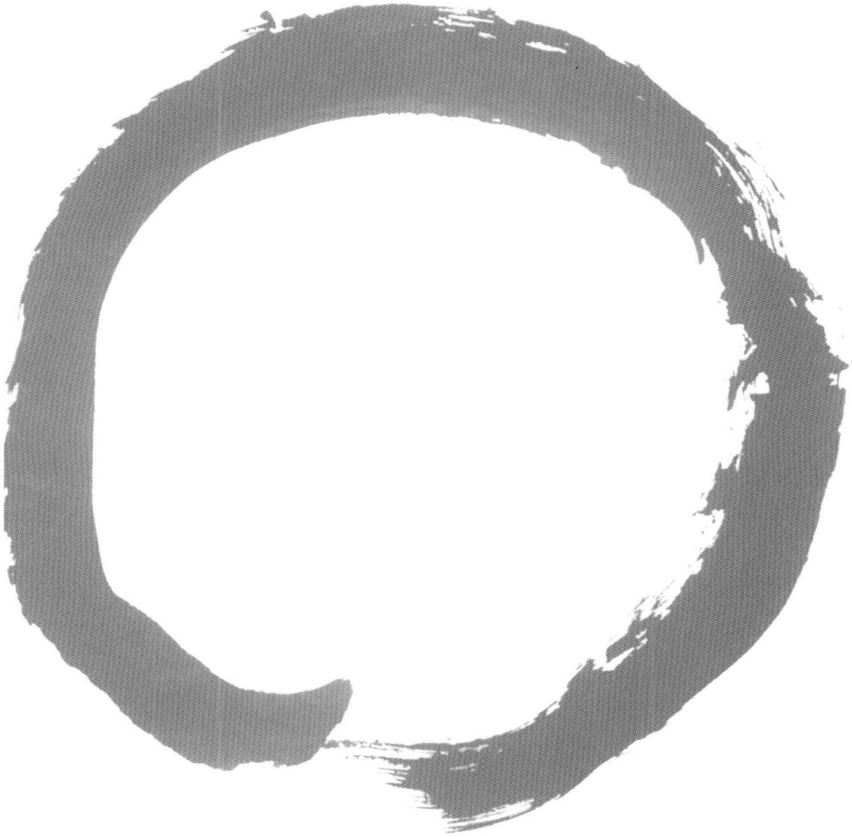

Contents

Legacy Forward

Chungwei Su
Principal, J. J. Pan and Partners

It was 1976 when Mr. Joshua Pan, who was amongst those few Western educated and trained elite architects from Taiwan, decided to relocate his practice from the well established professional environment of the United States back to a then under-developed Taiwanese market. The post WW II economic boom attracted worldwide architectural scholars and developers to major American universities and cities during the 1960s and 1970s. To most young architects from Taiwan, the privilege of attending top US universities for education and then practicing in renowned offices has been a guaranteed successful career model during that period of time. Nevertheless, by settling down within a comfort zone and striving for a firm partnership in a foreign land also meant breaking away from the motherland and leaving the nourishment of one's native culture. In spite of all the risks, Mr. Pan's decision to return with what he had learned and began the endeavor to build up a healthy professional environment for his homeland was clear.

Defying the under-developed architectural practice environment of Taiwan in the 1970s, a lot of Western trained architects started their offices with new design thinking coupled with solid knowledge and experiences, which initiated a wave of refreshing development that lasted through the next few decades. However, these success stories gradually dissipated after decades due to the lack of core values, firm heritage and management necessary for sustainable generational transitions.

As an architect, it is quite natural to pick a well established market with thriving opportunities rather than to enter a vague system full of uncertainties. However, for Mr. Pan, the choice to return meant that he was determined to fully utilize what he learned and experienced through 12 years of study and working in the United States to bring about a more mature practice environment to the land where he grew up. It takes a highly matured professional with confidence and clear cultural recognition to choose this uncertain path. For him, the basic belief of architecture behind his pursuit of excellence falls on the harmony with nature, the proper application of technologies and materials, and sincereness towards users. These principles established by Joshua Pan not only have guided the firm for the past four decades but will continue to guide JJP's future generations through various challenges and lead to a glorious future.

承 · 传

苏重威
潘冀联合建筑师事务所　主持人

潘冀建筑师 1976 年回到台湾，至今已逾 40 年，可算是相当早期的返台学人。回顾 20 世纪现代建筑风起云涌的 70 年代，由于亚欧大陆“二战”后的经济重建，使得当时的美国大城市与建筑学校成为全球建筑精英汇聚的殿堂。台湾战后经济条件欠佳，对于许多建筑新人来说，进入美国顶尖学府、大型国际事务所并奠定专业基础，已成为留学海外的成功模式。然而留在异乡安逸地执业、成为外国事务所合伙人，也意味着失去了一展抱负，为自己的故土提升建筑环境的机会。

相对于成长快速、制度成熟的西方社会，1970 年代的台湾建筑环境有待改进，专业制度也未臻完善，却仍有不少留学海外的建筑前辈选择了返台创办事务所，发挥所学，也带回建筑新思维。而今回首，曾经风光一时的建筑师们，能清楚而完整地建立事务所核心价值，并将理念传承制度化，继而永续经营的却不多见。身为建筑师，在专业与机会都成熟的环境执业，相对容易且自然。然而潘先生却在 40 年前选择了一条不一样的道路：在完全融入美国文化后却决定回到成长的地方，试图将国外所学的专业与制度，努力实践在自己的土地与文化上。走一条完全未知的路，必然需要在专业上有高度的成熟与自信，热爱并认同自己的文化，且对社会有强烈的使命感，而这些清楚的认知即是建立在对建筑的基本信念上：合乎自然与功能的法则，正确使用技术与材料，关怀环境与社会，坚守人本价值，并不断追求卓越。如同日月星辰之于海上行舟，一旦定位，航行者就能有所依循地向目标前进；40 年的基本信念也将如建筑航道上的坐标，继续引领着 JJP 的新世代，坚定不移地驶向未来。

01 LEGACY

Enlightened by modern architecture's emphasis on professionalism and discipline while guided by a clear design philosophy and purpose on the path of practice

启蒙于现代建筑的专业及严谨务实精神，传承理念、态度与设计意涵的实践之路

传承

Ethos

Set your heart upon the way

- respect nature

Base yourself on integrity

- harmony with the environment

Lean upon goodness and virtue

- compassion for man

Enjoy free spirit in the arts

- freedom in design

Verse 6, Book 7 of *The Analects of Confucius*

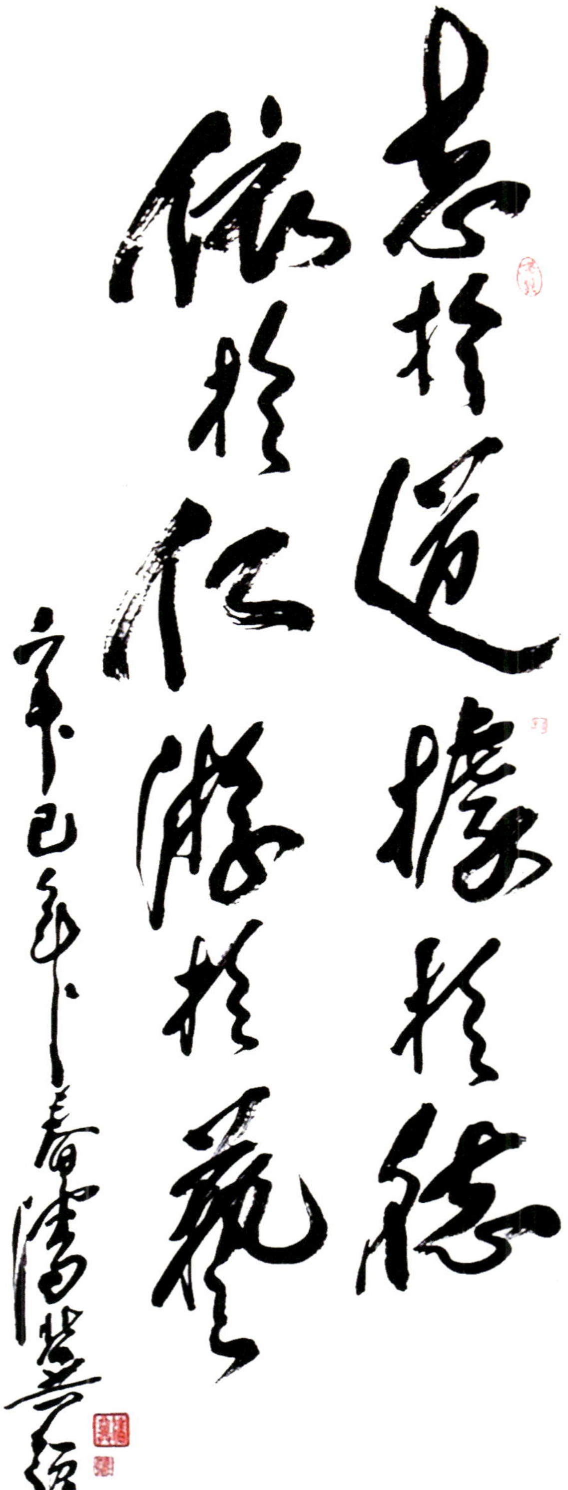
志於道據於德
依於仁游於藝
辛巳年春

1960

Foundation 建筑启蒙

1

Architectural Education

In 1964, after graduating from Taiwan Cheng Kung University and completing a year of military service, I continued my architectural education in the United States, where I studied at Rice University and Columbia University. Soon I came to the realization that the difference between Eastern and Western cultures was also reflected in their respective styles of education. Back then in Taiwan, students' final designs were graded by teachers in the library behind closed doors and with no chance for explanation. In the United States, I found that final designs must be presented on stage. Which meant that a student needed to figure out a better design process to arrive at a solution and then convey the design intention by developing presentation skills. Furthermore, the Taiwanese architectural curriculum placed a heavy emphasis on technical skills, which was quite different from the United States. For example, at Rice University, I took courses in art history, psychology, logic, sociology, among other subjects in addition to the core architectural courses. I recall that period of time to be quite tough both physically and spiritually, yet it became the most unforgettable experience of my studies. It was through art history that I understood the historical evolution of art and its context in relationship to architecture. The study of psychology helped me to be more aware of users' requirements so that I could develop a more reasonable architectural framework. The study of logic trained me to arrive at solutions to problems by thinking about the underlying causes. Sociology allowed me to understand the social environment and behavioral background as times progressed. Together these courses formed a solid foundation for my career in architectural design.

"Architectural design should not be confined to the ivory tower, but should serve the under-privileged classes of society in order to help them gain a better life."

“建筑设计不应该关在象牙塔里闭门造车，而要利用专业能力来帮助社会各阶层的人，让他们获得更多的幸福。”

After completing my studies at Rice, I enrolled in the Urban Design Institute at Columbia University. Urban design explores the interaction between the urban environment and the public, whereas architectural design focuses on the interpretation of a building on a single site. I studied under Professor Percival Goodman, who early in his career had co-written *Communitas: Means of Livelihood and Ways of Life*, a very famous and enlightening work that dealt with modern architecture and social ideas. As an architect, Professor Goodman had very strong views about society. He rarely talked about architecture in general; instead, he would discuss topics such as urban renewal and design issues facing the South Bronx in New York City, in an attempt to explore how to change social behavior through urban design. This course had quite an impact on me for it provided a deeper understanding of diversity in American society. Architectural design should not be confined to the ivory tower, but should serve the under-privileged classes of society in order to help them gain a better life.

建筑教育历程

台湾成功大学建筑系毕业后服役，1964 年赴美留学，先后进入美国莱斯大学、哥伦比亚大学接受西方的建筑教育，我意识到东西文化的差异也反映在教育的思维上。早年台湾的建筑系评图，老师是把所有人的设计图贴在图书室的墙上，关起门来打分数，没有任何说明的机会。赴美后，才知道所有的准备必须通过上台来讲解，于是我开始思考如何为自己的设计找出更好的方法来传递思考过程与设计意图，一步一步培养表达的能力。另外，台湾课程设计偏重工程方面，以莱斯大学的求学经验为例，除建筑课程外，亦修习艺术史、心理学、逻辑学、社会学、美国历史等科目，那是一段精神与体力都十分沉重，却是最难忘的读书经历：艺术史让我明晰历代艺术的演变，领会与建筑结合的脉络；心理学有助于理解使用者的需求，架构出合理的建筑空间；逻辑学训练因果关联，追查问题的始末，找到解决之道；社会学则是了解时空变迁中的社会环境与行为背景，所有的课程对我日后的建筑设计都有极大帮助。

1 Studio review at Rice University
设计课评图，莱斯大学时期，1966

2

3

4

Foundation for Architectural Practice

In the next ten years after my graduation from Columbia, I worked at three world renowned architectural firms successively: Philip Johnson & Richard Foster, Davis Brody and Associates, and Collins Uhl Hoisington Anderson.

During my time at Philip Johnson, I took part in two projects at Washington Square for New York University. One of the projects was the Bobst Library, the exterior wall of which employed a sand stone exquisitely made with modern techniques. The project

之后进入哥伦比亚大学读城市设计研究生，城市设计与之前所学的建筑设计的最大差异，在于前者较为注重单一基地的诠释，而城市设计则要深入探讨城市、环境跟人群之间的关联与互动。当时师从普西沃·古德曼教授，他在早年有一本十分有名的著作《社群乌托邦》，算是结合现代建筑与社会思想的启蒙书籍。教授是一位社会观念很强的建筑师，一整年的设计课程不谈常规建筑，而是讨论诸如纽约市南布朗克斯区的城市更新与设计议题，讨论如何透过城市设计的手法来改变社会行为。这个课程对我的冲击不小，让我对美国的多元社会有更深入的体验与了解，建筑设计不应该关在象牙塔里闭门造车，而要利用专业能力来帮助社会各阶层的人，让他们获得更多的幸福。

建筑师的基本功

哥伦比亚大学硕士毕业后，随即进入三家国际知名的建筑师事务所工作：菲利普·约翰逊和理查德·福斯特建筑事务所、戴维斯·布洛迪建筑师事务所及柯林·悠·霍伊辛顿·安德森建筑事务所，前后共约十年。

在菲利普·约翰逊和理查德·福斯特建筑事务所期间，我参与了纽约大学在纽约华盛顿广场的两个项目，其中博斯特图书馆外墙是以现代手法精雕细琢的砂石墙面，单单门框就 600 多种不同材料的门框节点，在施工图上就需交待清楚。在美国，施工方要把建筑师的设计想法转成一套现场可据以操作的图面，建筑师必须审查、核对、最后核准，我的工作是图面核对及确认。核对那些施工图原本是一件很无聊、单调的事，不过，我当时却觉得有趣，就像是看武林秘笈，去搞懂一套图面的前因后果，了解所有设计的精髓所在。

之后，来到戴维斯·布洛迪事务所，他们曾设计 1970 年大阪世博会美国馆充气建筑，以及诸多独具创意的平价住宅及大学建筑。我选择去戴维斯·布洛迪的原因，是他们的建筑设计师可以一路从设计构思、施工图到施工阶段的检视，除此之外，戴维斯·布洛迪事务所致力于改善城市关系，重视人际间的互动，在有限的预算下，努力将空间及材料运用到最好。由于认同他们的理念，我在这里待了四年。

5

我首先参与水牛城分校的设计项目，这个新校区要在一片冬季严寒的沼泽地上规划6 000人使用的校舍，我提出以欧洲中古城市的配置概念为雏形，让建筑群体可以在寒冷气候中交互串联，活化校园的行为脉络。此方案被事务所选上，我这个来自台湾的东方人成为项目设计师，并负责整合其他建筑设计师的想法。秉持着努力做好每件事的决心，让我在上帝的保守下得到同事的信任与支持，一步一步走出一条路，一点一滴水到渠成，完成了水牛城分校的设计项目。

had more than 600 kinds of doorframes, plus countless materials, so the shop drawings needed to be very precise. In the United States, the construction company must develop shop drawings based on the architectural designs, and the architect must review and approve them prior to actual construction. Generally, checking shop drawings is considered boring and tedious, but at that time I found the task to be extremely rewarding for the drawings were like a collection of martial arts secrets to me. I took advantage of the opportunity to learn about the relationship between design and production drawings and how to architectural design is manifested.

In addition to doing exemplary work, such as the inflatable USA Pavilion at the 1970 Osaka Japan World Expo along with a number of creative affordable housing and university buildings, Davis Brody was also committed to improving urban relationships with an emphasis on interpersonal interaction, and making the best use of space and materials with a limited budget. The firm's working method also allowed architects in work on all project phases, from the original design then shop drawings, and all the way through inspection during the construction stage. Because of these common beliefs, I joined Davis Brody and worked there for four years.

The first project I participated in was the SUNY Buffalo Joseph Ellicott Campus. The project called for a new campus for 6,000 students built on a swamp in a cold winter climate. I proposed to use the configuration of a medieval European city as a prototype

2 Prof. Percival Goodman and his published *Communitas*
普西沃 · 古德曼教授与其著作《社群乌托邦》，1966–1967

3 3rd year studio works at Taiwan Cheng Kung University
成功大学时期建筑作品，1962

4 5th year studio works at Rice University
莱斯大学时期建筑作品，1965

5 SUNY Buffalo Joseph Ellicott Campus, Davis Brody and Associates
戴维斯 · 布洛迪建筑师事务所时期作品，纽约州立大学乔瑟夫 · 埃利科水牛城分校，1968–1972

6-A

6-B

之后，我进入位于新泽西州普林斯顿大学附近的柯林·悠·霍伊辛顿·安德森建筑事务所，他们需要有完整大型工程经验的建筑师，我为事务所架设了整个建筑设计与绘图制度。四年之中，我把部分欠缺的专业经验补足，除了建立设计流程之外，自己负责设计的项目一定全程参与，包括一般建筑师都不愿意接触的施工说明书和预算，这部分虽然有专人代劳，但我身为项目负责人，也应深入了解施工说明书、材料、造价等议题。施工阶段的工地周会一直由我主持，了解设计者与施工方的互动关系，对项目执行上的各项环节进行掌控，以如实还原设计本意。

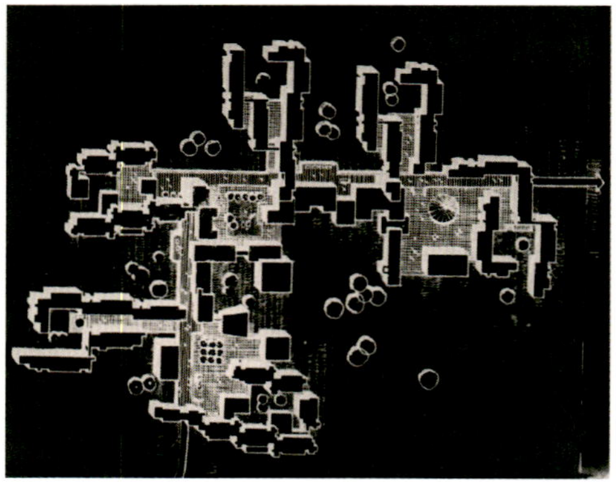

6-C

to interconnect the complex. My proposal was selected and I, as an Asian person from Taiwan, was appointed to be the project designer, responsible for integrating the various parts of the design. With a determination to excel and the blessing of God, I gained the trust and support of my colleagues. Step by step, detail by detail, the Buffalo campus project was successfully completed.

After Davis Brody, I joined Collins Uhl Hoisington Anderson, a firm based near Princeton University in New Jersey. At that time, they needed an architect with overall design experience on major projects, and I helped to set up the entire system of architectural design process and drawing system for the firm. I was also fully involved in the projects that I was responsible for, including the construction specifications and cost estimates, which are parts of the process that most architects tend to shy away from. Though typically a specialist would work on the specifications and estimates, I believe as the project architect, I should understand all aspects of the architectural practice, like specifications, materials, cost and other issues. In those four years, I certainly made up for some missing parts in my professional experience.

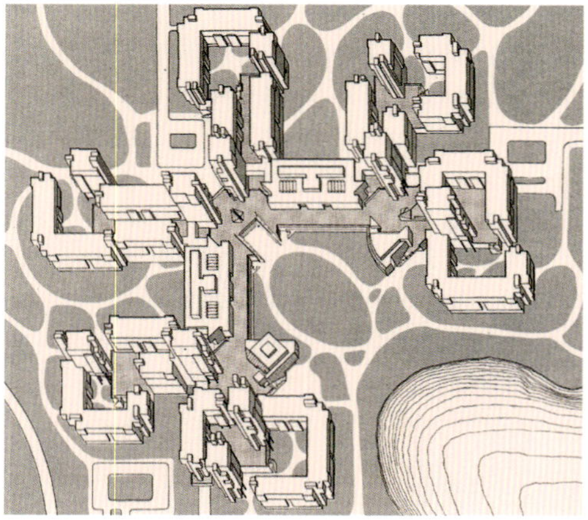

6-D

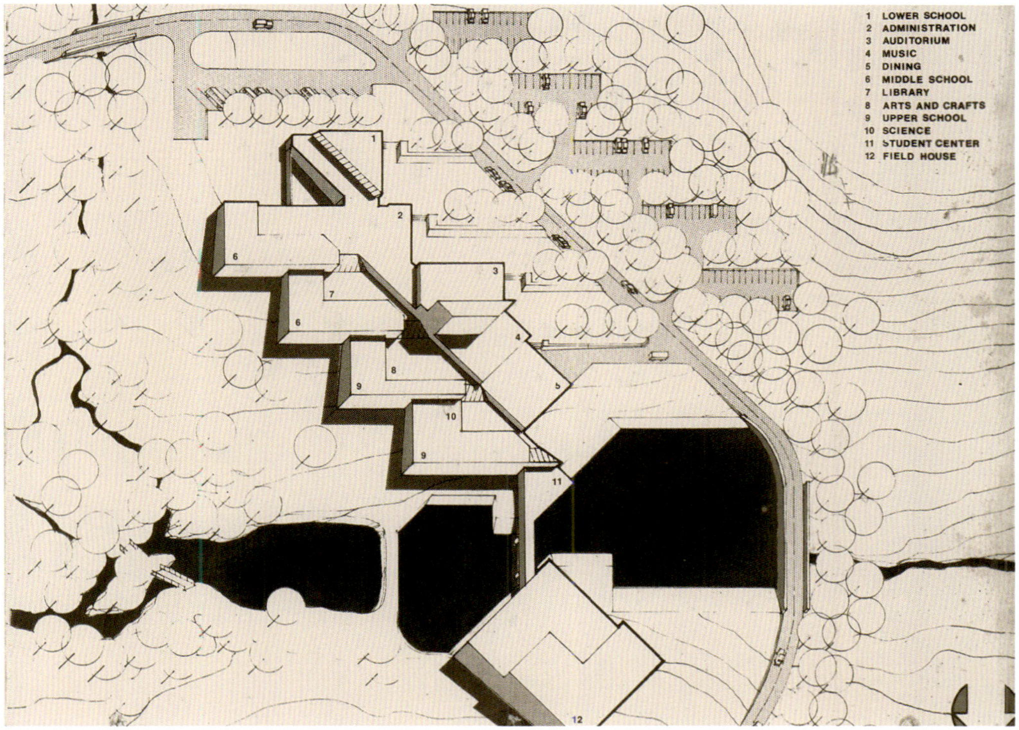

7

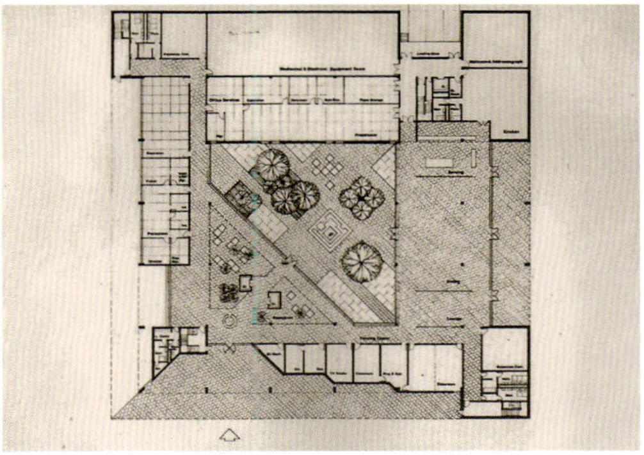

8-A

8-B

6 SUNY Buffalo, Joseph Ellicott Campus, Davis Brody and Associates
戴维斯·布洛迪建筑师事务所时期作品，纽约州立大学乔瑟夫·埃利科水牛城分校，1968–1972

A *Architectural Record* cover
美国《建筑实录》封面，1970

B *Progressive Architecture* cover
美国《进步建筑》封面，1975

C Conceptual sketch
概念构思手绘稿，1970

D Axonometric drawing
轴测图

7 New campus-Pingry/Kent Place school, Collins Uhl Hoisington Anderson Associate
平格瑞 / 肯特校园规划设计，柯林·悠·霍伊辛顿·安德森建筑事务所时期作品，1972–1976

8 Admin Building for Carter-Wallnce Inc., Collins Uhl Hoisington Anderson Associate
卡特·华勒斯公司行政中心，柯林·悠·霍伊辛顿·安德森建筑事务所时期作品，1972–1976

A Floor plan 平面图

B Model 模型

1981

Formation 自立门户

1

"It is difficult to change someone's mind, but I can change my own way of thinking so that I can step in the right direction."

" 虽然很难改变别人的偏见，但是，我可以改变自己的心境，往理想的方向前进。"

Return to Taiwan

In 1975, as the Vietnam War ended and Chiang Kai-shek passed away, Taiwan's society was in a transition. Invited by my old schoolmates, Mr. Philip T. C. Fei and Mr. Mei Cheng, I returned to Taiwan to compete for the design of the Chiang Kai-shek Memorial Hall and the National Theater & Concert Hall. By then, I had been working in the United States for more than 10 years and everything was going quite well. However, my wife and I both believed that we should eventually return and contribute to Taiwan's development, so we returned to Taiwan in 1976.

At that time, the level of construction companies and consultants was still rudimentary. Nevertheless, I insisted that all project-related items be clearly marked out. I devoted a lot of effort on coordination and integration, including the construction of the exterior walls, waterproofing, structure and mechanical. The attention to detail was such that I even drew ceiling plans myself and documented each switch and outlet position so that the space would be more convenient for the end users. The consultants couldn't understand my actions but I firmly believed architectural design should comprehensively address all aspects of a building instead of merely its appearance. Therefore, architectural designs must be feasible and practical in order to be carried out to the end.

However, such an attitude was far from the norm in those days. I also did not socialize, and disagreed with the way architecture was practiced. If I could not change such a rigid system, then who could I change? So for decades I chose to teach as a practitioner in the architecture departments of many universities. Since I had the fortune to experience the entire architectural process, whether in education or practice, I hoped to impart my own professional knowledge, skills, and experience to help young passionate students, rather than having that experience remain in the office. My goal was to disseminate the worthy concepts and nutrients I had acquired by means of lessons and lectures. I did so to help the students gain a more inclusive experience and adopt a more acute attention toward their own education, the environment and society, so that they can bring about positive changes in the future. I have held this philosophy to this day.

返台实践

1975 年美国从越南撤退，蒋中正逝世，台湾社会面临转型之际，我应学长费宗澄、陈迈先生之邀，回台湾参加 " 中正纪念堂及两厅院设计 " 竞图。当时已在美国工作十多年，一切顺利，无匮乏之虞，但是我和太太都有这样的想法——有朝一日我们应该回台湾贡献一己之力。慎重考虑之后，毅然举家于 1976 年返台。

回台初期，台湾施工方及配合顾问的水平都尚待提升，我坚持工程中所有建筑相关的项目都要清楚标示，花了许多心力在协调整合上，包括外墙施工、防水、结构、机电等，甚至亲自绘制天花平面，并要求明确标示每个房间的开关、插头位置，让建筑空间合乎使用者的行为。当时水电空调顾问都觉得不可思议，但我认为建筑设计是一门整体考虑的学问，对于建筑的照顾应该是全面性，而不是只把精神花费在讨喜的建筑表皮而已；因此，我对设计的态度是一定要能落实施工，并且如实执行到最后，这也是我一直秉持的实践原则。

1 J.J. Pan Architects
潘冀建筑师事务所，1981

Founding of J.J. Pan

In order to test whether my professional skills were adequate to support my beliefs, after four years of work at Fei & Cheng Associates, I decided to work independently in 1981. Looking back now I would say that I was quite young, and too idealistic and courageous at that time. In those days, people used to attain business through social relationships, and architects were no exception. However, I did not believe that was the only way to be an architect. I know it is difficult to change someone's mind, but I can change my own way of thinking so that I can step in the right direction. So I made the most important decision in my life, which was to found the architecture firm of J.J. Pan.

After founding the firm, I insisted to my colleagues not to make ethical compromises and instead to use our professional skills to convince others to do things in a fair manner. Even when faced with unreasonable regulations and officials who rigidly interpreted these rules, the firm would still present sensible arguments and base our persuasion by appealing to reason.

3

这样的态度难免与当时许多人不同，我不交际，也不认同当时台湾的职场文化，无法改变这个僵化的制度，我能改变谁？于是在工作的同时，我选择教书，并赴各大学建筑学系讲授设计，数十年如一日；我有幸接受到完整的建筑培养，无论是学校教育或是实务领域，我希望将自身累积的专业知识、技能与经验毫无保留地传授，造就更多有志于建筑的青年学子，而不是让这些经验只停留在事务所，更希望透过学校的教学与演讲，提供好的养分跟想法来影响学生或是周围的人，让他们对教育、环境、社会有更包容的体会和观察，进而改变社会。一直以来，我都是抱着这样的理念。

2

2 Group tutorial of Chung Yuan Christian University
中原大学设计课后讨论，1976

3 Office
办公室一隅，1981

创立事务所

于宗迈建筑事务所工作四年之后，为了测试自己的建筑专业能力是否可以坦荡地在社会立足，我毅然于 1981 年独立执业。现在想来，当时的勇气或许来自年少轻狂的理想性格，那时的台湾社会，一般人认为需有关系来取得业务，建筑师也不例外。只是，我不相信社会上所弥漫的风气是做建筑师的唯一方式，虽然很难改变别人的偏见，但是，我可以改变自己的心境，往理想的方向前进。于是，我做了这一生最关键的决定——创立潘冀建筑师事务所。

成立事务所之后，面对当时的社会现象，甚或听闻业界有些暗地里的不正当行为，我们都坚持不走后门也不妥协，仅以端正的专业态度耐心说理，尽力劝服对方，一同以光明正大之方式行事，对于不合理之法规及建管部门、人员刻板的释法方式，事务所亦能辨明清楚，据理力争。

4

5-B

5-C

Growing with Taiwan

In the 1980s, the real estate industry in Taiwan was booming and almost all architects were working in this sector. Although I had designed housing projects in the United States, along with academic and healthcare buildings. I found it much easier to communicate with clients from universities and healthcare organizations, so the Taiwan Ocean University became the first client of our firm. For the next nine years, the firm worked on more than ten buildings for the client, including renovations and new constructions. Although the cumulative budget was only less than five hundred million NT dollars, our efforts allowed the Taiwan Ocean University to form a unified campus, and also laid the foundation for

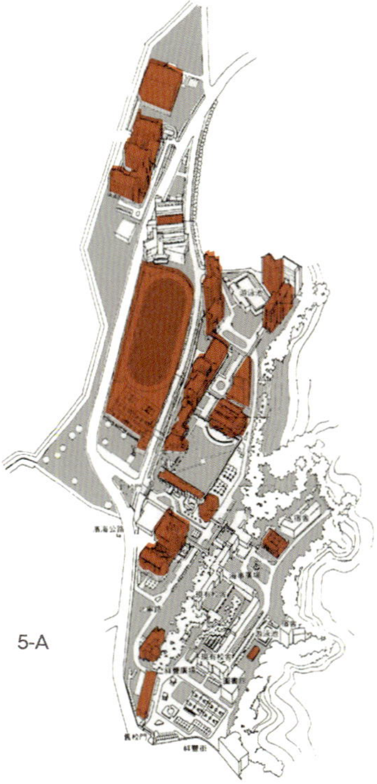

5-A

4 Members of J.J. Pan Architects
事务所初创成员，1981

5 Multiple-Stage Renewal, Taiwan Ocean University
海洋大学校园规划设计，1981–1990

A Axonometric drawing of campus 轴测图

B Before 规划前

C After 改建后

our firm's expertise in campus building design.

Around the same time, the Hsinchu Science and Industrial Park was established, so architects that could talk to R & D professionals were urgently needed. Starting from the design for the park's entrance gate, our firm began to continuously provide design services for the Science and Industrial Park and its tenants. The high-tech industry's requirement of tight environmental control meant its personnel must always be in an artificial environment and the project must meet stringent time and specification requirements while reflecting the appropriate corporate image. The firm's designs thus always seek to balance the functional aspects with our desire to create a humanized environment. In recent years, we have become even more responsive to environmental and ecological values, and adopted the principles of green building design

与台湾一同成长

1980年代，台湾房地产蓬勃起飞，所有的建筑师都往热门的方向跻身，而我在美国工作时虽也设计过住宅，但也有大学建筑、医疗研究设施等，回到台湾之后，我发现我与大学建筑、医疗研究机构的业主较能沟通，于是，海洋大学成了我开业后的第一个项目，包括旧建筑翻新、修改、兴建新建筑物等，九年间设计了十多栋建筑物，工程总经费虽然不到新台币五亿元，却让海洋大学呈现出整体大学校园风范，也奠定了事务所在设计校园建筑方面的基础。

6

7

8-A

8-B

to establish a new standard for sustainable industrial buildings.

As the firm's workload grew steadily, we did not forget our original intention of giving back to society. For example, Sung Chiang Poetry Park, an urban park completely supported by private donations and designed by our firm pro bono, was completed in 1991. On a visit today one can still get a glimpse of its humanistic style and enjoy it as a peaceful place for the soul.

Gradually, the projects designed by our firm began to receive more and more recognition and awards. In 1994, I sent the portfolio of our firm to the American Institute of Architects for review and I was honored to be selected as a Fellow of the American Institute of Architects (FAIA), which is the highest honor for American architects. This recognition proves that as long as you cultivate with diligence, no matter where you are, architectural design can always find its own path.

同一时期，新竹科学园区成立，需要能与专业科技研发人士对话的建筑师，事务所从科学园区大门为始的设计业务，至今不断。由于高科技厂房讲究物理环境的严密管控，人员必须常处于全然的人工环境之中，设计除了需要精准响应整个制程规范及时程的要求之外，更需要量身定做科技办厂的企业形象，尤其着力在人性化的办公与休憩空间，不论室内外均设法提供一定的舒适度。近年来更逐步响应友善环境之生态价值，遵循绿色建筑设计规划原则，以期设立可持续建筑的新时代典范。

事务所的业务在稳定中成长，同时不曾忘却回馈社会的初衷，1991 年落成的松江诗园即为一例。这是一座完全由民间企业集资兴建的城市公园，并由事务所义务设计，直到今日，仍可一窥当年的人文风采，成为人们心灵休憩的场所。

事务所的设计作品逐渐累积口碑，屡屡获奖与受到肯定。1994 年，我将开业十余年间所设计兴建的台湾建筑作品送至美国建筑师协会审查，通过多层筛选，有幸获选为美国建筑师协会院士，得到美国建筑师的最高荣誉。这证明：建筑设计不管在何处立地，只要用心耕耘，都能找到属于自己的一条路。

9

6 First project at the Hsinchu Science Park
新竹科学园区首件项目，1981

7 Core team members
潘冀建筑师事务所核心成员，1992

8 Sung Chiang Poetry Park
松江诗园，1991

9 Honored as FAIA
获选为美国建筑师协会院士，1994

2000

Emergence 在地深耕

1

"We hope that this firm can be a positive force for society as well as the environment, and its design concepts can be constantly carried forward."

"希望努力创立的事务所可以是推动社会、环境向上的正面力量，设计理念也能不断传递延续。"

Transitioning to J.J. Pan and Partners

By 2000, our firm had grown to employ more than 150 staff. We had established standard operating procedures for every design process and construction administration, while the management of the firm had become more complete and mature. Ever since our firm was founded, the construction division strictly supervised construction sites in accordance with the specifications and quality standards, and to immediately reported back any deviations. In addition, within the first 24 hours after a natural disaster occurs, our staff will conduct an on-site inspection to report back to the office and keep our clients informed about the situation. This procedure has always been part of our high standards for project execution.

In Taiwan, when an architectural firm has grown to a certain size and age, usually the founder will gradually leave the office and the firm will eventually be dissolved. This means the waste of its accumulated resources which is a great pity, so I always had a clear vision to establish a firm with a sustainable inheritance. Fortunately, over the years the firm had accumulated many great colleagues that shared the same belief. Hence in 2000, J.J. Pan and Partners was formally founded to perpetuate the ethos of the firm. I hope that this firm can be a positive force for society as well as the environment, and its design concepts can be constantly carried forward.

The Challenge in Practice —— Innovation, Service and Execution

From day one the firm has attached great importance to every single part of the practice by adhering to three fundamental principles. Firstly, "meaningful and intrinsic innovation," which is to design architecture that is in harmony with space, time, and environment. Secondly, "dedicated and thoughtful service," which is to deliver results with foresight and comprehensiveness; Thirdly, "precise and reliable execution," which is to provide the best advice and solutions to our clients. By adhering to these principles, the firm has accumulated many long-term clients and partners who often invite us to expand our business. Consequently, from 2001 to 2005, we successively established the Horizon Design Shanghai and Xiamen offices to provide design consulting services in mainland China. The concept of innovation, service and execution has since been continuously promoted on both sides of the

改制联合

2000 年，事务所成员突破 150 人，每个建筑环节皆制定标准作业流程，各项系统及制度趋于完整且成熟；事务所开业以来即设立工务部，依据施工规范与品质保证标准，严格监督施工现场，遇有突发状况必主动汇报；另外，每次天灾发生后 24 小时之内，工地同仁会至现场检查后汇报事务所，同时主动通知业主，确实掌握施工质量。这一直是在项目执行上保持的高标准要求之一。

在台湾，通常建筑师事务所成长到了一定的规模与年份之后，创始建筑师会随岁月而退下职场，同时事务所也随之解散，形成社会资源的浪费，真是十分可惜。所以在我心中，建立传承的想法十分明确，而且事务所多年累积了许多资深且理念相同的伙伴，遂于 2000 年改制联合建筑师事务所，共同去维护这个建筑品牌，希望努力创立的事务所可以是推动社会、环境向上的正面力量，设计理念也能不断传递延续。

1 J.J. Pan and Partners
潘冀联合建筑师事务所，2000

2

3

2 Da Xia House, Chinese Culture University
台湾中国文化大学城区部大夏馆，2001

3 Taipei Truth Lutheran Church
台北真理堂，2004

Taiwan Strait which led to many awards both at home and abroad.

International Recognitions: the Da Xia House of the Chinese Culture University in Taipei was an adaptive reuse and expansion of an existing building. The project made use of the 15cm distance from the building line leftover by the original building to install a common skin that covered both the existing building and the expansion, as if it was a veil between the virtual world and reality that co-exists. This project won the AIA New York Chapter Annual Design Award in 2001 as the jurors noted that such a difficult urban problem was given fresh meaning through a simple design. This was also the first project from Taiwan to win the AIA New York Chapter Annual Design Award.

The Taipei Truth Lutheran Church, which won the AIA New York State Citation Award in 2006, was a breakthrough in the traditional impression of churches. Centrally located on Taipei's Xinsheng South Road, the project adopted the concept of being a spiritual guardian in the city and employed abstract and simplified design methods to express the various religious connotations. With a street-facing stair tower culminating in a glass light box on top adorned with a red cross, the church became a unique spiritual landmark in the city. The surrounding open spaces attract neighborhood residents to socialize. Although it took seven years from the building's initial proposal to final completion, the end result was highly praised by the church and the neighborhood, which deeply reflects how architecture can inspire its users and influence its surroundings.

Across the Strait in the Shanghai Cultural

实践的挑战——创新、服务、执行

在认真经营多年之后，适逢长期合作的业主及伙伴们提出邀约拓展大陆业务，于是在 2001—2005 年间，先后成立了上海群裕、厦门群道设计咨询公司。秉持“创新、服务、执行”的理念持续于两岸展开实践，作品开始在国际奖项竞技中崭露头角，受肯定的作品包括：位于台北的台湾中国文化大学城区部大夏馆增建项目，是旧建筑增建之案例。项目利用原建筑立面离建筑红线尚有 15 厘米距离，巧妙地以表皮的概念来包裹原有建筑物，隐约呈现原有建筑的面貌，如同虚实交错的面纱，并与新增部分连成一体，让新旧记忆的轨迹并存。本项目获得 2001 年美国纽约市建筑师协会建筑设计奖，评审一致认同透过简洁的设计处理，如此的城市难题被重新赋予了新意，这也是台湾作品首次入选该项国际大奖。

2006 年获得美国纽约州建筑师协会佳作奖的真理堂，则以城市心灵守护者为构思起点，突破了一般教堂给人的既定印象。项目位于台北市新生南路的城市核心区域，教堂以简化抽象的现代设计手法，传递出多层次的宗教内涵；面临道路的楼梯塔，顶端以玻璃光箱烘托出红色十字架，在城市中成为简洁静谧的精神性地标，外围留设的带状绿地与广场成为邻里居民流连互动的空间。这个看似光鲜亮丽的建筑，从最初提案到完成，历经七年的时间，当它落成献堂后，极受教会及周围小区好评，此时我更深刻地体会到，建筑可以感动人心，影响这个城市。

4-A

设于上海市杨浦区滨江文化创意园区的群裕办公室，基地曾是美国通用电气在亚洲最具规模的电子工厂，是极具价值的历史文化遗址。事务所租用部分厂房并加以改建，经由旧建筑再利用，在现代感与国际化的潮流中，融入本土地域的特色，以"建筑中的建筑"之形态交错配置其中，呈现跨时空的传承与延续，深化空间的历史意涵。本项目获得 2008 年美国纽约州建筑师协会杰出奖（旧建筑再利用）及 2008 年第二届美国《商业周刊》与《建筑实录》中国奖之最佳历史保护奖（后因 2011 年园区重建，群裕是年迁出该区）。

4 Horizon Design Shanghai Office
群裕设计咨询上海办公室，2005

Creative Industry Park, the site of the Horizon Design Shanghai office was once General Electric's largest factory in Asia. Inspired by the site's significant historical value, the design intention for the interior fit-out was to maintain the heritage of the building while integrating local and contemporary characteristics present in Shanghai. The spatial arrangement took the concept of a "building within a building" to create an interactive space similar to streets and squares. The project received the 2008 AIA New York State Award of Excellence for Adaptive Reuse and the Best Historic Preservation Project, and the 2nd Biennial *Business Week/ Architectural Record* China Awards in 2008.

4-B

2011

Culmination 续攀高峰

1

"It is fortunate that our firm has assembled a group of people who share the same beliefs and can work together for a common vision of architecture."

"很庆幸，事务所团队汇聚了一群有着相同理念的同事，为建筑实践的梦想而努力。"

30th Anniversary

By 2011, the firm had grown to employ more than 250 people. Combining the cultural inheritance and accumulated experience of our firm, we created a culture that could be characterized as "reflective yet expressive". The firm also began a tradition of commemorating every tenth anniversary. On the 10th anniversary, we had models of our major projects exhibited as a review of our completed and unfinished dreams. For the 20th anniversary in 2001, we invited the famous composer, Mr. Gordon Chin, to compose a score named "Vision of A Bridge". The theme of the music was to inspire respect and care for the environment. The premiere concert was held at the NTCH in Taipei with our clients and friends in attendance. For the 30th anniversary in 2011, I suggested to hold a student design competition. Since I was often invited to give lectures on both sides of the Taiwan Strait, I felt that young students should have more interaction and communication with each other. We also held an exhibition and a symposium as part of the event. The competition's theme of "Cocoon and Simplicity" focused on the issue of housing. As rising sea levels could cause flooding problems for major urban areas, so we wished to inspire creativity and imagination through the design contest. These commemorative events held every ten years are opportunities for our staff to cherish our achievements while fostering the motivation for moving forward.

Sharing the Achievement

The book, *Brick by Brick - Building the Foundation of Life*, was truly a beautiful surprise for me. After reading about our firm in magazines and mainstream media, a publisher thought that my personal experience might be inspiring for the younger generation. However, initially I declined to be interviewed, so the publisher invited Miss Goya Lan, a writer to interview many of the colleagues in our firm and my wife. Then the publisher asked me to confirm the content and to be interviewed many times myself. It took 2 years to share my personal principles and working experience before the book was finally published with my approval. For the book cover, the publisher found a poster of a collage, which was a gift from all my colleagues for my birthday in 2009.

游艺三十

2011 年，事务所两岸成员突破 250 人，加上事务所文化的传承与累积，造就了"不卑不亢、内蕴外扬"的团队特色。我们每十年都会举办周年活动，10 周年时，我们将 10 年的经典作品模型展示在事务所，重新检视我们完成与未完成的梦想；2001 年 20 周年时，特别委托邀请作曲家金希文作了一首曲子，取名为"梦见一座桥"，音乐会在台湾中正纪念堂戏剧院举办，邀请了事务所的业主和朋友，因为我们希望透过音乐，提醒大家当时还很少人谈到对建筑与环境的尊重与关怀；2011 年 30 周年时，除了在台湾举办活动之外，因常受邀两岸演讲，有感于青年学子应多方交流接触，就提议举办两岸学生比图，也为竞赛成果筹划展览与座谈，甚至发行专刊，主题为"茧与简"，主要探讨居住问题；近年来大家都意识到海平面上升会造成陆地受淹，衍生出各类型的居住变化形式，希望借着比图提供建筑人脑力激荡的园地。从每十周年的活动中，让事务所的全体同仁一同回顾与检视，珍惜现有的成果，成为继续前进的动力。

1 J.J. Pan and Partners
潘冀联合建筑师事务所，2011

The collage consists of each colleague's photo, indicating that everyone has different backgrounds and characters, but we can all work together and contribute our own strengths to achieve great architectural projects. It is fortunate that our firm has assembled a group of people who share the same beliefs and can work together for a common vision of architecture.

In 2016, I was awarded the Taiwan Awards for Arts, the highest honor afforded to an artist in Taiwan. The award is not just an honor for me, but also a recognition of our firm and a representation of society's expectations for architectural design. Architecture is a complex process which cannot be completed independently, so it is necessary to have a team to work together on projects that exhibit concern for the environment and society. Through the Taiwan Awards of Art, I called on society to pay more attention to architects and have their projects reflect the enduring values of society, history, culture, and the environment.

2

3

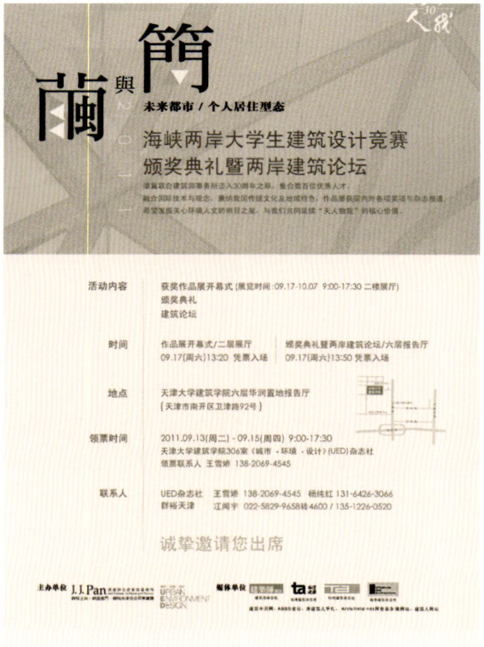

4

2 10th anniversary exhibition
10 周年作品展，1991

3 20th anniversary multimedia concert
20 周年多媒体音乐会，2001

4 30th anniversary cross-strait student design competition
30 周年海峡两岸学生竞赛，2011

5

6

建筑成果的分享

《人生基本功》这本书对我而言是一件美丽的意外，出版社在媒体、杂志上看到事务所过去的报道，认为我的个人成长历程，对年轻人具有启发性，但我一直婉拒。后来出版社邀请作家蓝丽娟小姐，访问事务所内诸多同事及我太太，最后再要求我确认内容，并又进行了多次访谈，历经两年用文字叙述方式与读者分享我在事务所的做事原则与工作经验，最后经我同意后集结成册。出版前夕，出版社发现一幅海报拼图，这是在我 2009 年生日时，同事合送的礼物，海报由每位同事的照片拼接而成，深具意义，代表每个人虽然有着不同的背景与特质，却能相聚在事务所发挥专长，成就建筑作品。很庆幸，事务所团队汇聚了一群有着相同理念的同事，为建筑实践的梦想而努力；而《人生基本功》这件意外也因读者们的持续回响，名列诚品书店的年度选书单上。

Creator of Green Buildings

In 2013, we cooperated with the Delta Electronics Foundation to design the "Ring of Celestial Bliss" as the main lantern for the Taiwan Lantern Festival. The concept was derived from the Chinese phrase "Blessings as high as the sky", and was conveyed through a circular band of projected images and LED lights as metaphor for nature's perpetual cycle of life. Further inspired by the historical and cultural characteristics of Hsinchu, the structure employed all reusable materials. Steel was selected as the primary structural system complemented by bamboo cladding in a shifting inclination to create a more dynamic visual effect. Appearing as a glowing object hovering in the night, the project represented profound respect for nature while also providing an auspicious vision for the future. The project won the 1st Prize of the *Global Views Monthly* CSR Award for Environmental Protection in 2013, and the Jury Award and Popular Choice Award in the Architizer A+ Awards (Pop-up & Temporary Category) in 2015.

Our firm made environmentally friendly design a core value long before the popularization of green buildings. At the Main Library of Chung Yuan Christian University in 1985, we had already began to employ design methods for shading, heat insulation, energy saving and natural ventilation. With the advancement of green building related technologies, we now have more professional tools and resources to go even further.

5 Book cover, *Brick by Brick - Building the Foundation of Life*
《人生基本功：建筑师潘冀的砌砖哲学》封面，2011

6 Ring of Celestial Bliss
永续之环，2013

7

8

2016 年获颁台湾文艺奖，这不仅代表个人的荣誉，更代表对事务所的肯定及整体社会对建筑设计的重视与期待。建筑实践是一件结合众多专业的复杂过程，无法一个人独立完成，需要志同道合的伙伴。事务所得以共同达成许多令人引以为傲的目标及作品，共同实现珍惜土地及资源的谦冲情怀，透过这个奖项，我更呼吁社会重视建筑师的价值，让建筑作品反映出社会、历史、文化、环境共荣的可持续价值。

绿色行动的时代见证

2013 年，事务所借由台达电子文教基金会的跨界合作，为台湾灯会设计了“永续之环”——一座以绿色建筑手法设计的主灯，运用生生不息的“环”来诠释中国人“鸿福齐天”的意念，结合新竹古称“竹堑”的地域特色，采用可回收再利用之材料将竹子、帆布、钢结构经由建筑工法，表现出前倾后仰的渐变结构，悬挑漂浮于夜空之中，内部以最新投影技术及 LED 灯光创造出 270°的环场影片，传达尊重自然与挑战未来的思维，这件艺术作品的展出时间仅十五天，吸引了大批台湾内外报刊杂志报导。建筑认同没有地域区别，“永续之环”以文化内涵为构思起点，传达尊重自然与挑战未来的意念。本项目获得 2015 年 Architizer A+“瞬间快闪类”建筑设计评审奖及观众票选奖，及 2013 年《远见》杂志“企业社会责任奖”环境保护组首奖。

7 Taiwan Awards of Art
获颁“台湾文艺奖”, 2016

8 Architizer A+ Awards gala
Architizer A+ 颁奖典礼，2015

9

长久以来，事务所以友善自然环境为设计之核心价值，早在绿色建筑概念普及之前，1985 年落成之张静愚纪念图书馆，即以遮阳、隔热、节能与自然通风的设计手法来达到功效。随着节能工法与技术的演进，事务所有了更专业的工具与资源。台达台南二期厂房巧妙融入屋顶花园、绿化休憩阳台、浮力通风演讲厅、生态池、混合密林等，获得 2013 年台湾优良绿建筑设计奖；台积电中科十五厂区更以立体交迭绿化、中庭节能设计、水资源生态沟渠、动物生态廊道、智能控制系统等，入围 2014 年 WAF 世界建筑节再生能源及回收利用项目奖项，同时获得台湾优良绿建筑设计奖；另一件海外作品台达美洲总部大楼，取得了美国绿色建筑 LEED 最高等级（白金级）认证，成为美国弗利蒙市的绿色新地标及当地首座预期实现 " 净零耗能 " 愿景的建筑，同时也获得加州伯克利大学人造环境中心宜居建筑佳作奖。2015 年底应台达之邀，在巴黎联合国气候变化大会（COP21）上，分享台湾在地绿色建筑经验，希望借此加强对人造环境的意识，为这片美好的环境与土地作出贡献。

The Tainan Plant Phase II of Delta Electronics ingeniously integrated the roof garden, leisure balcony, ecological pond and a mixed jungle garden into the design and was awarded with an Outstanding Green Building Design Award in 2013. With the incorporation of features such as a greenbelt, an energy-saving atrium, wildlife crossing underpasses, and an intelligent building control system, the Fab 15 plant of Taiwan Semiconductor Manufacturing Company was nominated for the Production Energy and Recycling Award in the 2014 World Architecture Festival. Phase I of the project was also awarded the Outstanding Green Building Design Award. The headquarters for Delta Americas, completed in 2015, has become a green landmark of Fremont, California. As the first local building with the aim of achieving "Net Zero Energy," the project has already obtained LEED Platinum certification and was awarded the Livable Building Award by the Center for Built Environment of UC Berkeley in 2017.

Lastly, in December of 2015, I was invited by Delta Electronics to attend the Paris Climate Conference and shared our experience in designing green buildings and addressing environmental concerns in Taiwan. I sincerely hope that we can count the continuous blessings in our lives and contribute to repairing and restoring the environment.

10

9 Speech at COP 21 Paris
巴黎联合国气候变化大会演讲，2015

10 Delta Americas Headquarters
台达美洲总部大楼，2015

future

Review and Renew 继往开来

1

"In the future, JJP as a family will continue to cultivate compassion as it explores all design possibilities on the path towards realizing a more humanistic architecture."

"未来，属于新世代的 JJP 团队将持续以人本为念想，尝试设计的各项可能，为作品写下新页。"

The first three internationally published monographs of JJP each covered five to ten years of our works in those periods. As the firm heads into its 37th year of existence, the mission of this fourth book is to serve as a retrospective of the firm's work to date as well as the inclusion of some recent projects.

While reviewing past work, invariably there are mixed feelings along with many blessings. Judging from contemporary standards, I believe many of the projects could have been done better, but in retrospect, the most conspicuous accomplishment of our firm is the brave acceptance of new challenges. By adopting a learning while doing approach, the firm has completed multiple unprecedented design projects in Taiwan.

In terms of building type, there were research facilities for biological high molecular, nano and photon technology; along with the most advanced production plants for the IT, IC, semiconductor and TFT/LCD industries. Furthermore, there were healthcare facilities for radiation, proton, cancer hospitals and elderly and dementia care, as well as state of the art aquariums. All of the above projects were considered pioneers in Taiwan.

In terms of building technology, the firm was an early adopter of steel construction, drywall, metal and glass curtain walls, space frames, dual-curved long-span structures and has completed several internationally recognized green architecture projects. Besides the technical aspects of these accomplishments, the firm injected cultural, social and environmental empathies in giving these projects a humanistic dimension both within and without.

While service and delivery are aspects where JJP has long exceled, not for a moment has the firm forgotten that the difference between "building" and architecture lies in creativity and innovation. Such is especially true in the age of artificial intelligence when architects can be easily surpassed and replaced by robots.

Facing an ever-changing future, what adjustments should be made by this firm with nearly four decades of experience and over 300 professionals?

The following measures are what we have committed ourselves to:

执业之路走来栖栖遑遑，兢兢业业，竟也三十有七，以人生周期来算，正值青壮之年，但回到创意为主导的职业规划，却已届传承之际。时值第四本作品专集问世，除了分享近年新作，也重新检视前三本作品专集成果，借由回顾三十多年的创作及执业历程，省思过往，定调未来。

细数前作，有感恩也有感慨，许多早期作品以现在的设计标准来看，虽都有可改进的空间，但也确立了事务所最主要的成绩及对社会的贡献，在于不畏挑战、边做边学地承接各项首遇的设计任务。建筑类型从生物科技、高分子、纳米、光子源同步辐射加速器等实验设施，以及集成电路、液晶平板显示器等高端先进工业建筑，扩展到放射、质子、癌症医疗、赡养失智人群，以及领先国际的大型水族馆设施等，都是台湾首创案例。技术层面更涵盖了干式施工（钢结构、内墙、帷幕）、三度空间桁架、超大跨距、高规绿色建筑等范围，都大胆尝试并成功达成，也始终坚信在生硬的"技术"任务中，注入人文、社会与环境等关怀，是温润建筑的最佳养分。

1 J.J. Pan and Partners leadership
From left: Hsiao-Ming Chang, Ya-Tian Chuang, Jason Chen, Chungwei Su, Joshua Jih Pan, Chung-Tsai Huang, Jong-Yu Cheng, Robin C.K. Tang
潘冀联合建筑师事务所核心团队，2017
左起：张晓鸣、庄雅典、陈洁生、苏重威、潘冀、黄种财、郑荣裕、唐正国

1. Re-emphasize the importance of design and innovation while recruiting more young and creative new blood into the firm.
2. Become more selective by scrutinizing potential new projects for design opportunities. Avoid over-reliance on past experience and working procedures, which has lead our colleagues to fall into the "comfort zone" trap.
3. To establish a "JJP Lab" to serve as a "think tank" with the charge of looking beyond the typical design process, and to tackle new issues, new challenges, new technologies that our profession is facing this day and age.
4. To better utilize JJP Shanghai's qualification as the first ever independently registered Local Design Institute (LDI) in Mainland China from Taiwan, as well as the first firm with professional qualifications in Mainland China, Taiwan and the US, in order to meet the increasingly complex market conditions.
5. To undertake more international projects than before and to collaborate with leading foreign architects on the global stage.

I have had the good fortune to contribute to the society, the environment and mankind through over forty years of professional practice. In the future, JJP as a family will continue to cultivate compassion as it explores all design possibilities on the path towards realizing a more humanistic architecture.

多年来，即使在服务与执行层面的完成度上累积佳评，但“回归人性的创新”始终是事务所的首要课题，少了创新、人性，建筑物仅是没有生命的硬件，尤其在人工智能大跃进的时代，仅着眼在造型、炫技的建筑设计是很容易被取代的，更容易忽略建筑的出发点在于创造美好的生活与环境。

面对世代快速接替的未来，这累积了近四十年经验、规模近三百人的事务所，该如何调整步伐？经过“承前”的思考与检讨，我们以更具前瞻“启后”的措施期许未来。

1. 强调创新设计的再精进，力求思维、手法、策略同步创新，放眼国际，吸引相同理念的年轻世代加入。
2. 慎选具创意挑战之项目，不以既有口碑为喜，在稳健的基础上开创新局。
3. 强化智库“JJP Lab”的实质内涵，深入探讨每个设计元素，积极发掘并面对现今时代可能产生的挑战与议题。
4. 成立中国唯一具有中国、美国以及中国台湾地区三方设计资质的“上海潘冀建筑设计事务所”，迎向更广阔、更多元的市场机制。
5. 争取更多境外设计机会，贡献具影响力的作品。

过去，我有幸以专业贡献所学，为人群，为环境，更为社会，四十年来的所学所得尽付书中篇章；未来，属于新世代的 JJP 团队将持续以人本为念想，尝试设计的各项可能，为作品写下新页。

2

3

4

5

6-A

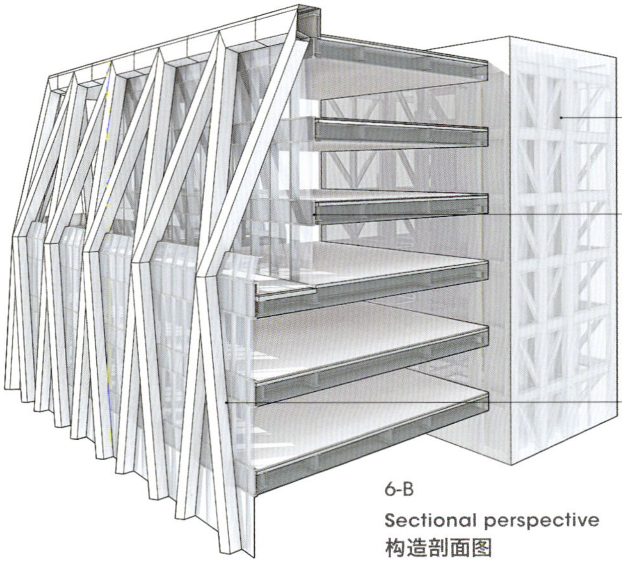

6-B

Sectional perspective

构造剖面图

2 China Wheat Products Research & Development Institute
台湾中华谷类食品研究所，1984

3 Winbond IC Wafer Fab II
华邦科技竹科二厂，1992

4 Quanta TFT/LCD Plant
广辉电子液晶薄膜显示器厂房，2001

5 Gymnasium, Chinese Culture University
台湾中国文化大学体育馆，2005

6 Taipei SCSB Headquarters
台北上海商银总部大楼，2017

Acknowledgements 致谢

Architectural practice is not an individual creative pursuit but rather a collaboration of various disciplines. Each project would not have been possible without the extraordinary contributions of our clients and collaborators, who trusted us, challenged us, and sustained our goals. Special thanks to all past and present JJP colleagues whose professionalism and dedication over the years are clearly reflected in these pages.

建筑作品不是单个设计者所能完成。JJP 感谢每一位业主与协力合作的伙伴们，每一次的信任与托付，激励与肯定，甚至挑战与试炼，都使我们成长跃进。每件作品更是所有同仁坚持与投入的成果，谨借一页名录，感谢所有潘冀人！

1981 | 潘冀 陈洁生 王才茂 谭俊发 张全庆 林纯莲 刘正帆 孙崇逸 | 1982 | 郑荣裕 郑泽雄 何银泉 张义震 陈金珠 | 1983 | 吴建辉 庄朖 许国 | 1984 |
周大同 廖伟立 | 1985 | 张芬梅 陈连盛 | 1986 | 蔡恒升 江建平 | 1987 | 黄种财 程顺德 黄永彬 王继志 朱有三 黄智隆 | 1988 | 黄天浩 黄富华 王凯宏 贺照梅
郭俊良 柳春堂 孙义德 雷昭子 汪胜元 | 1989 | 刘淑贞 陈慧秋 谢惠忠 张国兴 曹延筱 张庭柯 卢昭宏 陈宪德 房瑞芬 张弘昌 吴富美 黄舜君 简穗智 王锦光
徐木根 | 1990 | 黄丽明 施铭泉 林淑美 程怡华 李孟崇 江圣立 周美君 张俊玉 苏怡樵 李素美 吴佳惠 顾逸人 黄木锦 苏重威 吴方诚 | 1991 | 叶皖镇 游毅文
蔡志伟 张玮育 吴俊龙 黄正宏 洪玉琴 程家骥 陈敦欣 谢宏嶽 郭义斌 张庭栋 吴成化 李文山 | 1992 | 张豪德 丁玲 林俊伟 张志明 简雪芬 朱维瑾 张丽玟 游顺益
陈文扬 洪宏达 赵珮诗 李彦杰 黄志龙 吴淑精 张卫中 孙立采 王子亦 李志城 王昭婷 李永康 赖宗慎 陈君慧 傅碧铃 何颉 钟立亮 郑顺益 | 1993 | 蔡祖烈
刘信 邱智勇 王曼君 林文泽 刘子仪 廖珮雯 李春生 连思恩 黄文辉 李俊仪 杨文通 张莉红 江四福 林镱汰 林翠芬 | 1994 | 杨清祺 陈绍平 关惠尤 陈永昌 李皇辉
曾文忠 叶明明 叶中平 古春林 黄琮钦 刘戴智 邱德祥 陈以仑 钱云儒 梁钟远 张玮育 汪家璋 唐正国 李珮珮 | 1995 | 许泽志 邓志雄 冯祥云 黄翠琳 张忠琳
关念琪 苏明修 李志铭 吴淑慧 朱宏灏 胡汉凇 谢宗芳 | 1996 | 王慧仪 DON K. MARCHILDON 蒙旦光 黄瑞民 邱昭呈 程宏德 邱瑛瑛 梁宏达 张伟诚 王嘉珍
| 1997 | 赵化宇 陈伟民 翁晓兰 陈永云 严纬勇 陈伯桢 栗正暐 殷姿吟 许嘉麟 杨曼青 张芳瑜 余秋华 苏建纹 黄瑞孜 程宗贤 张柏豪 萧倩玉 | 1998 | 黄介帅
陈缵华 徐伟绩 黄广祺 陈莉华 王秋娟 林淑惠 陈隆 黄亦青 高睿康 舒哲伦 林坤志 李两成 杨淳媖 廖志坚 黄俊昇 许龙辉 高兴康 | 1999 | 孙健行 史晓梅 徐健铭
黄彦霖 陈取竫 刘俊龙 赖敏政 阙河彬 简巧欣 吴柏德 毛以励 黄家辉 黄驯明 王秋斌 沈庆豊 | 2000 | 施大伟 张元琳 苏亦奇 黄世铭 方凯平 潘硕 洪四娘
侯秉宜 张家绮 王铭远 郭玮克 曹国威 郭云莲 邓家文 游铂益 陈庆融 郑兆宏 宋契德 程露华 庞贝柔 林庆庭 尤彦斌 陈彦如 诸汉益 杨忠荣 陈雅祯 田佳弘 吴振兴
蔡政颖 许佳阳 陈上元 陈怡贤 钟明衡 刘志仁 詹幸惠 吴仁惠 李昱霖 林洵可 张柏彤 周太平 林清文 | 2001 | 黄严仕 董言斌 蔡斌君 徐晓文 张虹 张萌 田正
顾爱群 董桂芹 | 2002 | 葛兴光 黄文萦 侯冠如 林正侃 林永琪 林崑炤 李翩 陈慧珊 李立凯 蔡昇宏 谢伟士 陈启祥 陈周舆 高慈麟 卢玠玲 Martin Hagel 叶仲立
龚国彰 许弘明 李维迥 李宗荫 顾错 贾春光 范凤翔 杨琦 陶文彦 龚轶川 朱建忠 吴旻 | 2003 | 邱钦洲 杨宙燕 张宏森 SHIGETAKA ITO 李鸿勋 周正峰 简昌学
王文棋 陈松如 陈奕豪 罗淑娥 郑威玺 吴顺郎 刘家麟 杨士正 郑仁杰 严永森 陈嘉芸 陈永华 王承恩 李沅和 卢威宇 谢依璇 沈建宏 陈南宏 徐舒浩 吴璟 胡雄伟
李鳌 兰宇 唐海霞 韩涛 张园园 刘晓亮 刘春梅 张威 刘瑞婕 潘凌 高鑫 | 2004 | 杨天阔 陈重仁 赖仁达 李志坚 陈瑞雯 翁崇兴 杨高明 卢瑞恒 蔡大仁 高志伟
洪亮辰 林佳纬 谢寿明 陈景仁 刘亿松 钟易佑 何宗阳 曹国龙 吴常庆 方文言 林子莉 陈迪华 郭乃仁 吴轩铭 罗文琍 张淮萱 陈展裕 刘宜珮 王文錤 杨恩达
沉千量 黄丽华 郑宇宏 蔡慈媛 陈逸文 刘建生 吴昱廷 林鸿伟 何明强 吴征石 傅雪蓉 陆侃 丁志东 张力文 余俊宏 马梓兼 徐传宗 赵才先 井晓黎 方勇 | 2005 |
李佳玲 陈建宇 张旂彰 黄建祥 张晓鸣 吴晓云 宋伟祥 张懿芬 吴姿桦 毛语薇 翁嘉懋 陈晓彤 覃亦仙 林仲雍 施钧凯 周娟 陈亮 许怀利 顾辉 | 2006 | 吕理俊
李三民 黄诗芳 曾柏庭 柯炳荣 戴慧如 谢宗晏 颜嫣 梁建元 熊博远 苏美 郭容恕 苏政宏 彭琦瑄 陈燕 汤文胜 况伟 林飞燕 周炳杉 刘胜志 张简淑美 刘亚
金曜栋 顾金华 李富强 朱曌臣 吕迪 方丁云 周 玮 王晓琼 王小东 | 2007 | 冯如筠 薛文华 丁家伟 李月乔 曾彦智 田诒铭 曾能芳 吴宛飔 陈昆丰 吴德建
李志彦 钟欣锜 黄巾倪 陈米山 王文生 杨燕 徐程明 陈翰申 张莹 许丽珠 程昊 王友胜 李珺 汤富智 黄俊宪 朱恩进 陈柏宏 李嘉峰 曹丹哲 卢泰安 林圣平
张芳尧 袁芳中 张耿宪 聂赛男 高禹 洪珩元 许舒乔 江闻宇 陈孟杰 张洁华 李清汉 叶胜钿 詹佳芳 许裴耘 庄凌云 | 2008 | 林阿妹 蔡铭晃 钱信辉 刘峻志 王思敏
谢佳融 陈敬鹏 邱建渊 洪亮 张志贤 侯洸伍 詹谨维 王家珍 周鉴钟 张永秋 张韶芳 洪怡姗 潘永庆 刘钦福 林双楠 刘占庆 李贞仪 陈惠琪 廖纪勋 潘天壹 李欣屏
王素娟 秦亮 林志翰 黄韦仑 史伟富 张智钦 高卫 岳少龙 姜智匀 林圣芬 徐国浩 薛峰 陈志荣 朱普亮 杨琇茹 廖俊华 余志东 赵丽茹 朱虹 | 2009 | 翁源彬
林柏阳 陈佳伶 王蜀鹏 刘仁刚 王勇 黄跃华 魏嘉成 潘岩 黄圣闵 梁世伟 杨念恒 黄建智 林文森 陈建霖 李世民 林泓宇 沈瑞武 罗俊昇 廖哲逸 吴明辉 李铭洲
| 2010 | 黄钰珊 王永军 杨儒宪 黄天佐 蔡佳宏 陈聪贤 蔡尚益 黄再文 孟磊 刘为杰 吴玮梓 张宸样 吕小东 张仲宏 游静 束仕华 凌琳 傅泓文 谢绵丽 高成志
严丹璞 易人杰 孙璐娜 刘彦宏 徐圣运 陈鑫翰 梁向宇 毛轩扬 汪柏成 周丽芬 李琛 高铭伟 高幸萱 邓柏欣 刘长兴 许堡朝 邢茜 陈政宏 黄建中 颜佑展 唐明宗
吴柏璋 高钱钱 梁君汎 胡楠 傅君琪 赵爱丽 杨家琳 陈武志 蔡霖菁 林育纬 刘彩霞 林鼎翰 | 2011 | 王钟敏 许智程 吕俐蓁 杨镇瑞 黄伯翰 倪兵 魏鸿林 吕国维
叶宗舜 李大威 黄松柏 孙芳燕 许凤琴 王子齐 林显洋 贾朝阳 毛云霞 林禹成 张晏维 许浚齐 林立人 许凯铭 廖超逸 陈瑞源 王雨 何昌东 宋海洋 魏世英 王伦
高尚 林怡德 郭孟旻 柳昶 沈心婕 林颀轩 陈昭邑 袁以真 佟鹏 陈彦璋 | 2012 | 郤东禹 陈奕元 许舒涵 苏玮 袁薇薇 钟声 陈扬升 李家豪 朱育萱 游明宪 王金祥
王腾弘 黄婉婷 吴昊明 黄璿羽 李怡丽 陈镱铚 魏绍翔 梁廷达 刘晓霞 康兆言 杨家铭 徐培乔 马虹 | 2013 | 程安志 苏圣喆 黄英婷 魏名妤 徐岩奇 欧阳超
夏润钢 张钧婷 刘冠良 洪嘉谕 肖鑫丰 黄巧颖 庄巧彬 马威 廖惟宇 钟安媛 陆颖 陈子卿 简良霖 吴佩仪 | 2014 | 欧阳钧 黄明聪 郑渤如 杨智杰 廖汉伦 王朝庆
黄登懋 许芳语 罗娴 陈诗芸 徐钰 颜洪胤 陈彦希 陈钰铃 林建松 林继勇 林尹羚 林立婷 周晓艳 李思聪 冷雪霜 王金昌 王超 张采璇 | 2015 | 李瑞渝 杨青烨
谢维伦 陈诗禹 林紫滢 李伟汉 陈良慈 张哲望 李得嘉 颜伯菘 冯佳惠 陈怡伶 彭柏铮 程航 赖奕学 刘天龙 陈盈洁 唐伟哲 张博隆 许旭辉 黄建旺 鞠晓强 郭建兵
谢易珊 黄纪棠 Rafael Martinez 陈韦纶 张丝甜 陈柏瑞 吴致纬 宋士奇 余启佑 谢静 林柏峰 蔡琬琳 张佩珊 李孝文 杨子莹 祝艺苗 毛映壹 王志铭 蔡毓琳 王建成 辛丹
魏于婷 陈映全 | 2016 | 潘永询 陈信彦 许胜富 庄子毅 何睿奇 孙建能 夏亮 邱鸿来 叶政晖 张强 赖则安 杨雪 朴政泰 何玮玲 简佳禹 彭珞寒 霍旭亮 郑智谦
| 2017 | 严诗婷 黄盈慈 郑雅雯 吴昌佑 陈亭均 赖俊雄 邱敏 林晖杰 廖国威 沈可航 蓝德武 罗鸿远 范舒雅 许亚芸 陈颖奕 苏楠泽 洪培恩 邵晴 宋修慈 宋泓谕
萧钧豪 林育廷 邱瑀 叶馥瑢 杨平 翁思严 林宜陵 谢维伦 张惠宜 裴若安 吕纯云 陈璟纶 叶哲铭 洪士浩 吴友乔 蔡彦晖 陈彦呈 | 2018 | 庄雅典 田友虎 方小珍
王红瑞 樊凌云 王飞 刘朋 温鹤 陈倩 孙静 杨如意 简玉涵 柯佳君 侯仲楹 邱筱岚 柯采仪 林垂宾 蔡鸿志 蔡俊昌 杨世祺 何育宪 傅思蕙 杨智超 朱志恒 李政昇
徐政献 王楷婷 郑婷方 冯昱博 黎益伶 蒲骏逸 张峻玮 李政儒 苏峻弘 严洁 杨雅婷 胡淳 陈奂婷 姜晓恩 林晓韩 邱意淳 黄安泽 陈信颖 罗士洋

02 VISION

Break through the typical architectural constraints to imagine and explore future possibilities

突破常规建筑视角，探索并构筑创新未来的可能性

视角

DIS-COURSE

SHU-CHANG KUNG

The Inconstant and Constant in Architecture
from Service Creation, Innovation Drive to Paradigm Shift

龚书章

建筑的变与不变
——从服务创造、驱动创新到典范转型

Professor, Graduate Institute of Architecture, Hsinchu Chiao Tung University
Master of Design and Architecture, Graduate School of Design, Harvard University
新竹交通大学建筑研究所教授
美国哈佛大学研究所设计暨建筑硕士

Joshua Jih Pan, principal of J.J. Pan and Partners, Architects and Planners, returned to Taiwan from the U.S. in the late 1970s and founded his own firm in 1981. In over 30 years of practice since then, the overarching focus of Pan and the JJP team has been to integrate architectural technology and systems and to address concerns over society/culture/lifestyle on a foundation of professionalism, integrity and rigor. Pan and his team have always stressed that architecture must first and foremost embody "self-sufficient consistency" as a core principle. In addition to a well-coordinated design scheme as well as design team (including expert consultants, owners and the government), a consistent relationship between theoretical experiment and design realization must also be established. The same also pertains to internal consistency within the design concept and architectural systems, not to mention consistency in architectural realization and form and expression of the design values. As such, this principle is manifested in full across their diversified works, spanning from high-tech, religious, academic, medical and nursing buildings, to museums and campus planning. This oeuvre has framed for Taiwan a series of architectural paradigms that has and will continue to exert influence.

As Taiwan's society and industry undergo structural transformation over the past decade, Pan further endeavors to lead JJP to proactively respond to major challenges we face today with a more open and liberal position, while persisting with the constant values and principles of their professional practice. Such challenges include environmental and ecological concerns, climate change, renewable energy, industry transformation, aging society and low fertility. They strive to devise diversified and innovative approaches that at once safeguard cultural legacy and address sustainability, energy/carbon reduction and environmental protection.

From Service Creation to Innovation Drive

The designer "devises courses of action aimed at changing existing situations into preferred ones," said Herbert A. Simon, innovative thinker and Nobel laureate in Economic Sciences. The most charming aspect about this definition lies in its "optimism". He believes that the process of the human race responding to the challenge of the times for a better future is a "process of grand design". Meanwhile, none of the major problems humans face today falls under one single discipline; rather, they involve complicated social and technical systems and multiple stakeholders. Hence, design must turn from "creating objects" toward tackling these "grand problems" and "grand systems". One must break away from design conventions based on intuition and perception and, instead, explore a brand-new design culture.

In an age of global knowledge network and economy, the evolution and expansion of the design field is undergoing a paradigm shift from "creating style" to "driving innovation". On one hand, the role, mission, methodology and tools of "design" are changing in a drastic way, and the

"...architectural design has always been a way of thinking and acting, by which humans observe the world, discover and solve problems, create experiences and find values."

"……建筑设计一直就是一种观察世界、发现问题、解决问题、创造体验、建构价值的思维方式和行事方式。唯有有意识地面对问题的创新和创造，设计才会越来越接近其本来意义。"

JJP 团队主持建筑师潘冀从 1970 年代末留美归台，1981 年独立开业创办潘冀建筑师事务所至今，三十余年来一直以诚信严谨的专业精神，建筑技术与系统的整合理念，以及对社会 / 文化 / 生活的关怀，作为他和事务所团队最核心关注的焦点。潘冀和 JJP 总是以强调建筑首先必须体现其完整的真实性和一致性为其核心原则，不仅要求设计本身和设计团体（包括专业顾问、业主和政府机构）合作的协调性，也包括理论实验与作品实践的一致性关系。该原则也包括设计理念和建筑系统的内在一致性，以及建筑形式和基于设计的价值表现含义的一致性。因此，这一点在他们的众多作品中，如科技产业、宗教机构、学术机构、医疗养老设施、博物馆以及校园规划等，都有完整的具体体现，且为台湾建构出一系列影响时代的建筑典范类型。

而随着这十几年来台湾的社会和产业的结构性转变，潘冀更力图引领 JJP，一方面仍坚持其建筑专业秉持不变的价值和原则，而另一方面也继以更开放、更自由的姿态，来积极响应我们所面对的环境生态、气候变迁、能源再生、产业转型、高龄少子等议题的重大挑战，建构一个既关照文化传承，又应对环境节能减排、可持续思考的多元创新格局。

scope of design has also expanded to cover the design of relationships, interactions, services, systems, organizations and mechanisms. On the other hand, "design" has grown from a component of the industry chain and innovation chain to a contributor for entire systems and processes. The "designer" is becoming a co-creator of development strategies for society and industries and at the same time stepping out of the passive role of service provider to become a force that drives social innovation and spearheads a new wave of industrial revolution. Assuming a brand-new and more proactive stance, "design" is moving from a hired service in a vertically integrated system with greater industry potential value, to partially transforming into a collaborative partnership with future innovative industries on the creation of "knowledge capital", and further on to intervening in social and ecological environments.

With profound insights into these tremendous changes in our world, JJP has also embarked on a metamorphosis of its own while continuing to insist on a human-centered approach, team collaboration and comprehensive coordination and management, as it has over the decades. In recent years, JJP is beginning to shift from material design to creating experiences in non-material design, from integrated professional practice to trans-disciplinary innovation, from creative design to the marrying of design thinking and scientific thinking. They seek to expand the scope of design to cover the maximum number of people, propelling design into a bigger arena than ever. This new design thinking proposes higher and broader breakthrough innovations in the depth, breadth and integrity of knowledge, turning JJP into an idiosyncratic professional design team in Taiwan. Consequently, many of their recent works have also become powerful engines that continue to drive social and industry innovations in a changing world.

Consistent Principles and Diversified Eco-perspective

On the other hand, problems such as extreme weather, aging population/low fertility, economic inequality and energy shortage in a globalized world are leading us into unprecedented crises in a rapidly changing era. In face of these immense challenges, it has become ever more imperative and urgent to develop new technologies and systems to confront challenges from the real world and address the socio-cultural issues associated with the divergent transformation of complex societies.

Like many outstanding architectural practices of our times, JJP follows the design principles and thinking laid out by its founder. Its architectural designs embody sustainable concepts toward ecology and culture. On one hand, technical measures

Delta Americas Headquarters 台达美洲总部

从服务创造到驱动创新

诺贝尔经济学奖得主，也是富有创新精神的思想家赫伯特·西蒙曾经说过："所谓设计，就是一系列把现有情境往更好的方向引导的行为。"这个定义中最迷人的地方就是乐观。他认为人类应对时代挑战、探索更好的未来的过程，就是一个大设计的过程；而今天人类面临的主要问题，都不是单一学科的问题，而是涉及复杂的社会技术系统和诸多的利益关系人的问题。因此，设计需要从关注造物转而面向这些大问题和大系统；必须要超越以直觉和感知为主的设计传统，转而探索一种全新的设计文化。

在这个全球知识网络经济时代，设计的进化和设计领域的拓展，正在经历从创造风格到驱动创新的典范转型。一方面，设计的角色、使命、方法和工具正发生着巨大的变化，设计的对象也已经拓展到关系的设计、交互的设计、服务的设计、

are taken wherever possible to conserve energy and protect the environment by realizing diversified sustainability, and to make the ecological impact and energy systems of buildings important factors of consideration. On the other hand, a return to local culture and human-centered philosophy helps reconstruct historical memories for cultures and diversified psyches, thereby presenting a myriad of possibilities for people and their lifestyles.

"Glocal" Human-centered Sustainability

"Sustainability is design thinking that fundamentally responds to human life and the planet's ecology, and, more importantly, human-centered design that echoes the landscape and environment."

It is this at once local and global sensitivity to environmental sustainability and human-centered philosophy, coupled with an emphasis on innovative and interdisci-

Ring of Celestial Bliss 永续之环

系统的设计、组织的设计、机制的设计等；另一方面，设计从仅作为产业链和创新链的一个环节，拓展到了为系统和全流程做贡献。所谓设计者的角色，一方面正在成为社会和产业发展战略的协同建构者，另一方面也正在逐步走出被动的服务提供商的角色，成为推进社会创新和引领新一轮产业革命的动力。设计正在以一种全新的、更为主动的姿态，从激发产业潜在价值"一体化策略"提供服务的雇佣关系，开始部分转型成为与未来创新产业的知识资本的合作关系，进而介入改变社会和生态环境。

由于深切地了解世界发生了如此巨大的变化，JJP 数十年来，不仅坚守着对人性的理解、重视团队协作和完整协调管理，更在近年来全力从物质设计转向非物质设计的体验性创造，从专业整合执行转向跨领域的驱动创新，从创意设计转向设计思维和科技思维的整合等，将设计和多数人的未来联系在一起，从而使得设计进入前所未有的广阔领域。这种新的设计思维对知识的深度、广度和综合性提出了更高、更广的突破式创新，不仅使得 JJP 成为台湾别具一格的专业建筑设计团队，同时也使他们近年的许多作品成为时代变革趋势中持续驱动社会和产业创新的重要引擎之一。

原则的一致性和多元的生态环境观

另一方面，全球化下的气候极端变化、人口少子高龄化、经济不均危机、再生能源短缺等问题，使得我们这个快速变化的时代面临前所未有的危机。在此全球环境所面临的巨大挑战背景下，如何发展一套应对真实世界的挑战、针对复杂社会的多元转变，以及面对不断快速迭代的社会文化等问题的全新技术与系统，的确也越来越显示出其必要性和紧迫性。

JJP 也像当今许多杰出的建筑事务所一样，承续了创始人的一致性设计原则和思想，在建筑设计中体现出对生态和文化的永续概念，一方面将建筑的生态效应和能源系统作为重要的考虑因子，尽可能采用技术手段来节约能源、保护环境，实现多元且可持续发展；另一方面，则回到在地文化和人本思考，重新建构文化的历史记忆和多元精神，让这个社会面对未来充满了人和生活的种种可能。

既在地又全球的人本永续思考

"永续是根本地响应人类生活与地球生态环境的设计思维，更是响应地景和环境的人本设计。"这是潘冀建筑师一直以来面对永续环境的人本思考。

这个既在地又全球的主动积极的永续环境和人本思考，以及强调创新跨域的文化和科技共创，正好在 2013 年台湾灯会的台达"永续之

Taichung First High School Heritage Building 台中一中校史馆修复及再利用

plinary cultural and technical co-creation that led to a visionary milestone for JJP. The firm's contribution to the 2013 Taiwan Lantern Festival, entitled "Ring of Celestial Bliss" claimed both the Jury Award and Popular Vote at the 2015 A+ Awards. Furthermore, this "low-carbon Makino bamboo construction" served as a rich vernacular interpretation while successfully disseminated to the Taiwanese public an important sustainability concept that fuses "low-carbon architecture with energy-efficient technology".

The project uses systematic design thinking that respects nature and environmental sustainability and addresses resource shortage in both the global environment and its local rural setting. Meanwhile, the energy efficiency experiment featuring made-in-Taiwan technology underscores the beauty of Taiwan's culture and landscape while reminding the public to reflect on future climate and sustainability issues. The lantern festival, which is usually a convivial but resource-intensive annual event, is converted into an opportunity for participants to muse over

Taiwan Photon Source Facility 台湾光子源加速器实验中心

环”为 JJP 建构了一个面对未来的里程碑！这个备受重视的“永续之环”不仅获得了 2015 年 Architizer A+ 专业评审奖及观众票选奖双料殊荣，也以极具在地文化的桂竹低碳构筑，为台湾民众成功地传递了一个重要的以低碳建筑结合节能科技的永续概念。

这个设计一方面借以尊重自然、生生不息的永续环保的系统设计思考，来响应全球环境和在地偏乡的资源短缺，另一方面以台湾原创科技的节能实验来看见台湾地景文化的美，也提醒了民众面对未来气候的永续思考。整个作品让原本热闹但浪费资源的灯会，转变成为人们对环境永续问题的反思，进而在未来落实社会意识的改变！

JJP 也借这个作品的实验和实践，让自身的专业能进一步地扩展视野，让其建筑作品从环境到生活、从节能永续到文化再生，无论是“永续之

environmental sustainability and an agent of change for social awareness!

The experiment and realization of this project have also helped further broaden the professional horizon of JJP. The firm's diverse portfolio demonstrates consistent convictions and values about the environment, culture and sustainability--from environment to lifestyle, from energy conservation to cultural regeneration. Some emblematic concepts and projects include the idea of total utilization in Ring of Celestial Bliss, the high-tech net-zero-energy system in the Delta Americas Headquarters, the confluence of old and new in Taiwan Photon Source Facility, and the regeneration of historical memories in Taichung First High School Heritage Building. This trajectory of works helps power a kind of circular acceleration that changes our future civil society and ecological environment.

Systematic Magnification and Typological Innovation

For Pan, architectural design has always been a way of thinking and acting, by which humans observe the world, discover and solve problems, create experiences and find values. Only by consciously addressing problems through innovation and creativity can one approach the intrinsic meaning of design.Over the years, JJP has completed an impressive lineup of quintessential office buildings for Taiwan's high-tech companies. Not only did these projects attest to the leaping technological advances worldwide over the last three decades, they have also helped these companies fortify their brands, R&D capability, innovation, production and bespoke services.

Today, we are witnessing exponential global growth in next-generation information technologies, such as the Internet of things, mobile Internet, cloud computing, big data, remote sensing and geographic information system (GIS). At the same time, almost no city around the world is spared from staggering challenges and problems around climate, economic, social and political development; these issues are so daunting they cannot be tackled in isolation, but rather require global collaboration transcending national and disciplinary borders. The logistics industry around the world is also, as a result, facing transformation, and there is a shift from conventional plants to modernized infrastructure with innovation in energy efficiency, environmental protection and transportation in order to adapt to trends in customized service, economy of scale and clustering effect.

The Ally Logistics Yangmei Center, now under construction, is an illustrative example. Working together with Ally Logistic Property, JJP breaks away from the conventional and intuitive warehousing protocol and sets out to re-understand the fundamentals of communication, transportation and deployment between modern logistics and networks. The explorative experimentation and deduction through a series of spatial model studies led to an all-in-one intelligence-generating architectural prototype and solution for the logistics industry, max-

环”的完全利用、台达美洲总部的高科技／零耗能系统建构、台湾光子源加速器实验中心的新旧共生，或是台中一中校史馆的历史记忆再生等多元向度的永续思考，都建构在对环境和文化的永续信仰和价值之下，进行了一系列的环形加速运动，从而改变我们未来的市民社会和生态环境。

系统的放大，形态的创新

对主持建筑师潘冀而言，建筑设计一直就是一种观察世界、发现问题、解决问题、创造体验、建构价值的思维方式和行事方式；唯有有意识地面对问题去创新和创造，设计才会越来越接近其本来意义。

一直以来，JJP 为台湾的科技厂建立了一系列非常典范性的工业建筑，而这些工业建筑不仅见证了30多年来全球科技的大跃进发展和进步变迁，也为这些科技产业创造了品牌、研发、创新、生产、客制服务等多元典范。

如今，一方面由于全球的物联网、移动互联网、云计算、大数据、遥感遥测、空间地理信息技术等新一代信息技术的跃进发展，另一方面则由于全球城市的气候、经济、社会、政治的发展，几乎都共同面临巨大挑战和问题，因此需要跨越国界、跨越领域地全球性协作，无法单独应对。全世界的物流产业，也因此正面临转型阶段，逐渐从一般的厂房转化成在节能、环保、交通等领域都有创新思考的现代化物流基础设施，来顺应定制化服务、规模经济、群聚效应的趋势。目前施工中的永联杨梅物流中心，就是和永联物流开发共同摆脱传统直觉的仓储系统，重新就当代物流

Ally Logistics Yangmei Center 永联杨梅物流中心

imizing operational efficiency and minimizing operating cost.
The project applies the TOD (Transit Oriented Development) concept in an innovative way. A three-level warehousing system employing the "vertical stacking" idea in transportation infrastructure is incorporated in the plant, optimizing the loading function of the tri-story space. Both delivery efficiency and the flow of information and goods are also augmented. Various disciplines within the JJP team worked in a permeating way in probing the nature of "logistics" as an eco-system. They magnified "function" and "efficiency", two elements that have almost become routine in design. The magnification actually opened up the realm of design and endowed the conventional warehouse prototype with broader application and more freedom. The result is a "paradigm innovation" for the warehouse–from a traditional enclosed functional volume into a dynamic architectural landscape as an integral part of the environment.

It is exactly this kind of innovative thinking that has expanded "architectural design" from merely a component in the industry chain and innovation chain into something with far-reaching impact on the system, process and value chain of an entire industry. The scope of design has, in the process, also expanded from buildings to service systems and organizational mechanisms, effectuating a full paradigm shift from "creating style" to "driving innovation"!

From Protocol Experimentation to Value Co-creation

In recent years, swift changes toward an aging society and low fertility worldwide, and even in Taiwan, have made aging care, psychological healing and education for all ages into the most crucial issues and challenges of our times. Even though Europe or Japan have come up with some exemplary experimental models for care and services, they still continue to dedicate more effort into developing various design and innovation models for aging care, as they aim to find more precise angles with which to probe seniors' needs for physical and mental balance and resolve environmental integration problems under changing social structures. On one hand, they seek to help seniors achieve a sense of happiness and the society a sense of empathy and safety. On the other hand, they want to ensure more seniors in the future are able to maintain healthy relationships and connection with others and the world around them despite of their dwindling social links and physical capability and going through retirement, disability or loss of spouse. Therefore, further research, development and innovation should be done for our aging society into interpersonal and technology services that seniors would enjoy using, which will in turn help them maintain, reconstruct and ensure sound connection with the future society.
Pf. LIU Shyh-nan from the Institute of Creative Industries Design, Cheng Kung University made the following argument in his study

和网络之间的沟通、传递、整备等基础来进行理解、探索的实验和推演，试图通过尝试一系列的空间模型研究，找到物流工业一站式智能制造，达到最大化运作效率、最小化营运成本的建筑系统原型和解决方案。
此作品创新地运用 TOD（Transit Oriented Development）交通共构建筑的概念，为这个仓储建构了一个如交通建筑中转系统的“立体堆栈”三层仓储，既满足了三层空间卸货区的最大值，也让这个仓储如转运站一般提升运送效率，强化信息 / 货物流畅链接。JJP 团队通过不同专业工作领域相互渗透，借助对于“物流”这个生态系统的本质研究，进一步放大了设计当中经常惯性以待的“机能”和“效率”，反而因此扩展了设计的疆域，让传统的仓储类型有了更宽广的应用和自由，为仓储建筑开创出新的典范创新——从一个传统概念的封闭机能体量，成为一个环境当中具动态型态的地景建筑。
正因如此的创新思维，JJP 开始让建筑设计从仅作为产业链和创新链的一个环节，扩展到对整个产业系统、流程和价值链有了跃进式的影响；整个建筑进而随着设计的对象也拓展到所有服务系统和组织机制的设计，完整地达成从创造风格到驱动创新的典范转型！

从原型实验到价值共创

近年来无论是世界还是在台湾，人口高龄及少子化的急遽转变趋势，让赡养照护、心理疗愈、全龄教育成为我们最重要的课题和挑战！纵使欧洲或日本已经有部分的照顾服务实验模式傲视国际，但他们仍在不断试图从更精确的角度探测老年群体的身体 / 心理的平衡需求，以及社会结构转变的环境整合问题，试图研发出一系列有关老人照顾服务的设计创新模式，一方面找寻老人的生活幸福感、社会环境的同理心和安全感；另一方面则让未来更多的老人在历经退休、失能、丧偶等不断失去社会联结和行动能力之时，能够与人们和世界有好的关系联结。因此，如何开发研究更多让老人乐于使用的人际和科技服务，进而与未来社会维持、重建、确保联系，成为高龄社会一个研究创新主题。
成大创意产业设计研究所刘世南教授曾在他的研究《前辈的工作设计：高龄者的工作动机与工作生活》中提出：“由于平均寿命的延长、出生率的降低，以及战后婴儿潮世代届将集体退休，进入高龄社会已是全球性的趋势；而如何因应高龄社会的冲击，亦是各国社会政策与产业发展的重

entitled, "Work design for seniors: Work motivation and working life for seniors":

Aging society has become a global trend, as average longevity increases, fertility falls and post-war baby boomers retire collectively. How to adapt to the impact of an aging society will also become a critical issue for social policy-making and industry development around the world. To this end, the manpower policies undertaken by developed Western countries center on delaying retirement and encouraging re-employment to utilize senior manpower in a reasonable, friendly and effective way. "Reasonable" zeros in on justice in government policies and corporate hiring; "friendly" refers to providing supportive and assistive work environments and conditions; "effective" is about building a matchmaking mechanism for job supply and demand and job design that is conducive to work performance and sense of fulfillment...The topic of our times is how to enable "active aging" and "productive aging" in an aging society.

JJP has continued to address the greatest challenge facing our world today, whether by formulating the modern-day role of the church, helping to define the innovative spirit for schools, or further combining hospital and care institutions to create comprehensive elderly care systems and facilities. Specifically, how does one build architecture and environment for the future founded on the trinity of medicine, education and spirituality? How do we fulfill the needs of 90% of the population in the future to better achieve common good and social wellness through the creative vessels of social innovation and social design?

JJP's Social Welfare Park of Suang-Lien Presbyterian Church is an elderly care facility project that provides a healthy and interactive community with comprehensive medical services. Furthermore, it aims to create multi-layered and "public" social spaces alongside intimate cozy corners, by linking and unfolding interior/exterior open spaces. The tactic helps connect social communities while facilitating the learning

Suang-Lien Senior Living Community
双连教会社会福利园区

要课题。在欧美先进国家因应高龄社会的人力对策中，主要在于延后退休，促进再就业，追求合理、友善、有效地应用高龄人力资源。其中‘合理’讲究的是政府政策及企业雇用的正义性；‘友善’则在于提供支持与协助的工作环境与条件；‘有效’则在于建置就业供需的媒合机制，以及能够发挥工作绩效与工作满足的职务设计……当今的课题是因应高龄社会，如何创造出有活力的老年与有生产力的老年。”

JJP 一直以来无论在建构教会的当代角色、形塑学校的创新精神，或是更进一步地结合医院疗养机构来创造完整的老年赡养系统和设施方面，都持续地响应了当前世界所面临的最大挑战——如何在医疗、教育和精神的三位一体之中，完整建构一个面对未来的建筑和环境？如何能更进一步地透过社会创新、社会设计的专业创造，来满足未来 90% 的人的需求，以增进公共利益、社会福祉？

从双连教会社会福利园区作品当中，我们可以看到除了提供赡养中心一个健康互动且医疗照护完整的小区外，JJP 更试图一方面借由室内 / 室外开放的空间串联延展，创造多层次更具公共性的互动交际空间以及更具私密性的温暖角落，来回应生活社群的联系和生命的独处学习及回忆；另一方面，则进一步以“环境疗愈”的概念，如精神疗愈的礼拜堂、多元植物疗愈的花园、心理疗愈的故事记忆空间，以及物理疗愈的游走路径等，来创造一个为失智症长者服务的多层次单元照护环境。这个作品一再呈现以人为本的核心价值，不断地向自我专业提问并持续研究：银发族需要什么样的生活空间？他们可以负担什么样的照顾模式，才能让所有的老年人得到有尊严的照顾？

因此，这些年来除了建构一系列健康的老年赡养环境之外，更在这两三年企图面对社会未来更大的挑战——如何为更大族群的将近老年的熟龄世代，应对未来高龄社会来临所构成的生活结构转变，以及应对未来前瞻的科技发展可预见的新的生活机能，创造一系列更接近当今世代、更多元化的生活想象空间环境？于是 2016 年，在 JJP 的积极推动下，他们与勤美集团和璞真建设共同邀集了应用剧本实验室、自由空间教育基金会、林棋锦建筑师、谢伟士建筑师和我组成团队合作，以“同理心创新服务模式”为核心开始进行了“新世代熟龄好宅研究计划”。这个实验计划延续了 JJP 团队完整的专业经验，更进一步地以协同合作的方式，对熟龄世代通过一系列的情境仿真、体验共创、价值行为互动、脉络探询、并辅以如通用设计、赡养照护等相关议题，刺激出非常积

of living alone and living with memories. At the same time, a multi-layered environment of "unit care" is created for elderly dementia patients through the concept of "environmental healing"--a chapel for spiritual healing, a lush garden for phyto-healing, a story and memory space for psychological healing, and a meandering trail for physical healing. The core value of "human-centricity" resonates throughout the project, which tenaciously questions and tests each professional discipline and engages in research and development to answer these two questions: What kind of living space do seniors need? More importantly, what model of care can they afford so we can ensure all seniors can enjoy access to care and the dignity they deserve?

Consequently, JJP has committed itself to tackling a greater challenge in our future society in addition to designing an array of healthy environments for elderly care. That is, the challenge of how to sculpt more imaginative and diversified living spaces and environments for the even greater aging population in the imminent future while responding to changes in the "lifestyle structure" of an aged society and new "lifestyle functions" enabled by future technology advances? With this in mind, CMP Group and PUJEN Land Development, spurred by JJP, invited Scenario Lab, the Foundation of Universal Design Education, architects LIN Chi-Chin and HSIEH Wei-Shih and myself to form a team. With "empathetic innovative service model" at the core, we launched the "New-generation housing for the aging population (TBD/Next Generation of Senior Living)" research project. This experimental initiative picks up on JJP's well-rounded expertise and carries it forward in a collaborative way to galvanize progressive hidden insights and research methodologies through a series of scenario simulation, experience co-creation, value and behavioral interaction and context probe for the aging generation, supplemented by issues like universal design and continuing care. On one hand, the study aims to delineate critical values and potential new issues facing the future aged society and aged living. On the other hand, it helps designers appreciate the new "attitude to life" expressed by the aging generation arising from shifts in culture and values, as they develop strategies and architectural prototypes for the future. Some examples include mixed-age working communities/places (active and productive aging), "youth-and-senior co-living Airbnb", "senior co-operative co-living farm", "self-service" independent living (360-degree technology/telemedicine), diversified communication social groups/media, lifestyle/historical/artistic intervention (social participation through oral history/art), and co-sharing/co-farming edible landscape. It is readily evident that JJP is actively taking a great leap beyond their professional field to embrace this age of "value co-creation". They are endeavoring, through research insight and design innovation, to construct a prototype for "behavioral architecture" that allows the occupant's behavior and elements of the built environment to interact, evolve and

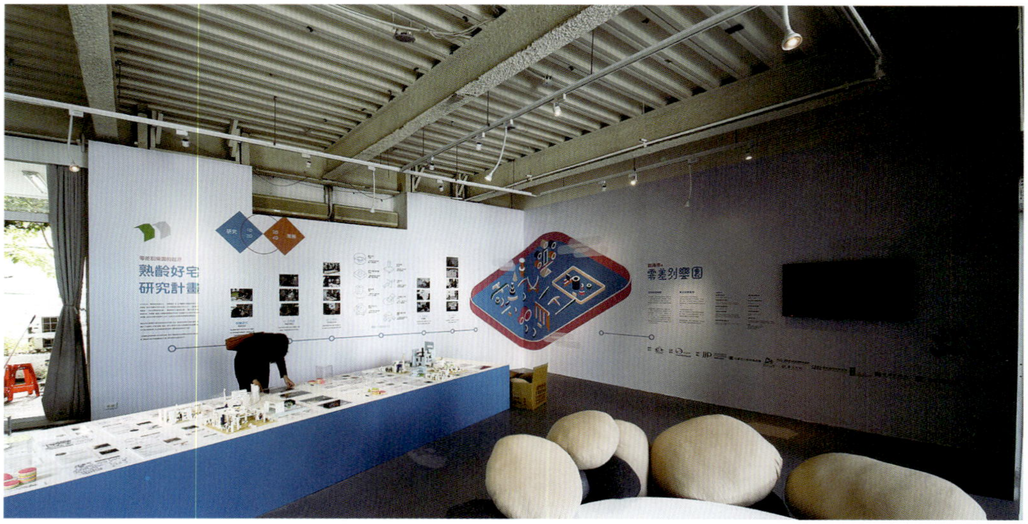

Interdisciplinary Workshop: Next Generation of Senior Living 新世代熟龄好宅研究计划

极的隐性洞见和研究方法：一方面勾勒出未来面对老年社会和生活的关键价值与潜在新议题；另一方面更协助设计者了解熟龄世代因文化价值变迁所表现出来的的新生活态度，来发展未来策略及建筑原型，例如混龄工作社群 / 场域（有活力、生产力的老年）、青银共居 Airbnb、熟龄互助同居生活农场、自我服务独立生活（全方位科技 / 远程医疗）、多元沟通社群 / 媒介、生活 / 历

calibrate in a sustainable fashion. As Pf. Liu fittingly puts it: "Aging (or growing old) is not tantamount to degeneration; as people age, their mental functions do not simply go through degeneration, but also develop mechanisms including growth, transformation and restructuring...One should resolutely set out to replace degenerative aging with active aging and consumptive aging with productive aging."

Toward a Paradigm Shift, Changing and Unchanging

In his 2017 Harvard commencement speech, Dr. Mark Zuckerberg's advices to the top university graduate in the world were as follows: "The challenge for our generation is creating a world where everyone has a sense of purpose...Redefining equality to give everyone the freedom they need to pursue purpose...It was never a question of if we were going to do this. The only question was when."
As I trace the journey taken by Pan and JJP over the last thirty-some years, I find them to be the epitome of the above philosophy. A team of architectural professionals focused on rigorous practice with bold and forward-looking mindsets, they are poised to go beyond their polished prowess as an architectural service provider and venture into the realm of "social design". They are working together with experts from different fields–including cultural, educational, social work, medical and technology–to communicate, pass on arts and cultural legacies, integrate technical resources and create common good, all toward the ultimate goal of a sustainable environment powered by "social innovation". I believe the JJP approach–transforming from "architectural design" to "social design" through changes in philosophical framework and systematic structure while insisting on changing and unchanging professional values–may help solve in a concrete way future sustainability issues for our society and environment. More importantly, this construct of social sensitivity and perspective will help shape vital core values as we face the future society, industry and even environment; thereby leading the architectural profession to take a giant step forward in a total paradigm shift.

Shu-Chang Kung
Taipei, Summer 2017

史 / 艺术介入（口述记忆 / 艺术的社交参与）、共享 / 共耕可食地景等重要趋势。由此，我们可以清楚地感受到他们正积极地从专业领域大步跨出，面向这个价值共创的时代，企图借由研究洞察和设计创新，建构出一个可以让生活者行为和建筑环境元素彼此影响互动、演化、多元永续调整的“行为式建筑”原型。就如刘世南教授的论述：“高龄（或称老化）不等于退化；随着年纪的增长，人们诸多心智功能不是单纯的退化，另外还有成长、转化、重组等发展机制……积极实践以有活力的老化取代失去活化的老化；以有生产的老化取代消费的老化。”

变与不变的典范转型

马克 · 扎克伯格曾在 2017 年哈佛大学毕业典礼上建议全世界最顶尖的年轻学子：“未来世界的创建，就是让所有人都具使命感的世界，以及重新定义平等，让每个人都有追求生活目的的自由！要不要这样做从来都不是问题，唯一的问题是什么时候去做？”
我清晰地看到潘冀和 JJP 在这三十多年一路走来，严谨实践且不断努力面向未来；他们企图由一个建筑设计专业服务者的角色，跨出自己已经成熟的专业领域，转而透过“社会设计”来和社会不同领域的专家，包括文化、教育、社工、医疗、科技等一起沟通、传承艺术文化、整合技术资源、创造公共利益、进而达到社会创新的永续环境。对我而言，JJP 这种思考框架和系统结构的转换、坚持变与不变的专业价值、从建筑设计转变成社会设计的工作方式，不仅有可能具体解决未来社会和环境的永续问题，更可以借助这种社会关怀视野的建构，成为我们面向未来社会、产业甚至环境的核心价值，进而为我们新的建筑专业做出新的典范！

龚书章
台北，二〇一七年夏

Library of Public Information
公共资讯图书馆

2007–2012
Taichung
41,797 m²

Stream of Knowledge in Digital Age
数字时代的知识串流

As Taiwan's first library of digital literature archive and multimedia, the Library of Public Information has transcended its traditional role of book-lending and become a center of leisure and boundless creative potential. Responding to the digital age, the library exemplifies the "stream of knowledge" connecting with the evolution of the urban environment to signify a new way of reading for the future.

公共资讯图书馆为台湾第一座数字文献典藏与媒体传播的图书馆，除了纸本馆藏资源外，更提供多媒体创作与共享学习的服务，以流动的内外空间形式来响应数字时代知识串流的现象，以可持续的思维创造城市与自然和谐共存的环境，以及属于未来市民的阅读新风景。

A Continuous Network of Public Art
连续性的城市公共艺术

With the façade featuring a "horizontal flow," the project seeks to both literally and figuratively acknowledge the urban history of Taichung, which began its development with the excavation of the irrigation canals during the city's early settlement. Situated between dense development areas and the urban greenbelt, the layout of the site and the building are designed to mediate the transition, with its main entrance fronting the green belt, easily accessible circulation paths, grand steps connecting the plaza and the central court embraced by the 'L' shaped building, establishing a continuous experience of urban public space.

台中市的城市兴起来自两条河流域间灌溉的沟渠，本设计建筑表皮的表情、环境光影的变化皆试图反映地域特性，让文化、产业、城市得以连续；图书馆犹如大型的公共艺术伫立于城市与绿化带两种不同环境之间，扮演重要的中介角色，因此以 L 形平面连接邻近的开放绿地，延续地景至公共广场，乃至建筑室内，建立连续性的城市公共空间体验。

Outside-in Reading Areas
内外感知的阅读空间

The strip windows of various heights and inclinations not only bring different amounts of daylight and outdoor views but also create various cozy reading environments for visitors. Such areas include wide expanses fitted with lounge chairs and reading tables, a bar table that overlooks the garden, a semi-outdoor story area, and a series of round skylights that allow glimpses of the shifting sky. Each floor's spatial arrangement, color and furniture design takes its cues from the corresponding outside view – events held outside, tree trunks, the tree canopy or crown, the city skyline, and clouds, fostering an outside-in effect. 1F Digital Lobby: The children's learning area employs a green canopy and animal theme with colorful elements such as bookshelves that recall hedges to create a spatial experience. 2F Multimedia: Vertical wooden arches and light pillars echo the tree trunks outside the windows. 3F Periodicals: The tree crowns are transformed into bookshelves and furniture for reading, creating an atmosphere of being in the treetops. 4F Science and Technology: Taking its cue from the urban skyline, the bookshelves are staggered in height and feature brightly painted, lighting-integrated display frames that evoke the city lights through the windows. 5F Humanities: The cloud-like curved white furniture echoes the blue sky above and is enhanced by the natural light filtered through the round skylights.

不同高低与倾斜度设计的水平带窗外墙，除引进不同的日光量与户外景色，也创造了各种阅读空间的可能性，包括临近宽阔景致的休闲桌椅及沙发、光线均匀稳定的实木阅览桌、远眺中庭的临窗吧台桌、亲近自然的半户外故事区，以及随天幕变色的圆形天窗等空间设计，提供不同个性的读者一方适情适性的阅读角落。以“借景”的设计概念将不同楼层窗景引入室内，转化为空间、家具、色彩系统等主题。各楼层皆定义为不同的空间属性：一楼为“社群集散场所”，儿童学习中心以森林与动物作为主题，塑造缤纷色彩的童趣世界；二楼为“树阵”媒体资源区，呼应窗外景象；三楼“树冠”是期刊报纸区，搭配专属桌椅、书架的友善阅读环境，塑造穿梭树梢般的空间体验；四楼“天际线”为科技产业阅览区，喻意科技如天弗届；五楼“浮云”为人文艺术阅览区，透过圆形天窗与自然天光，提升阅读乐趣。

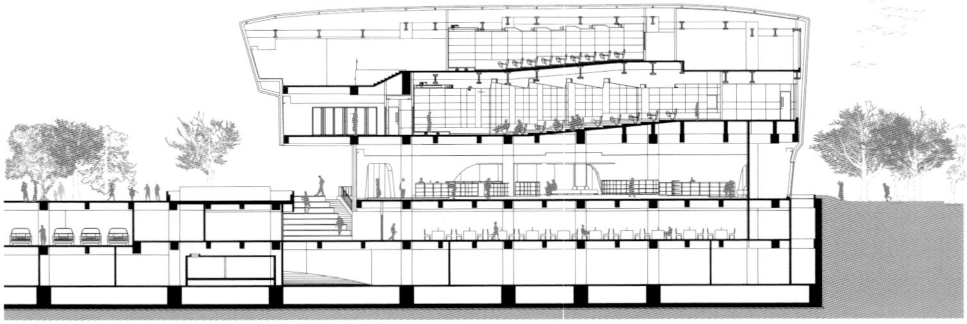

Section 剖面

0 10 20m

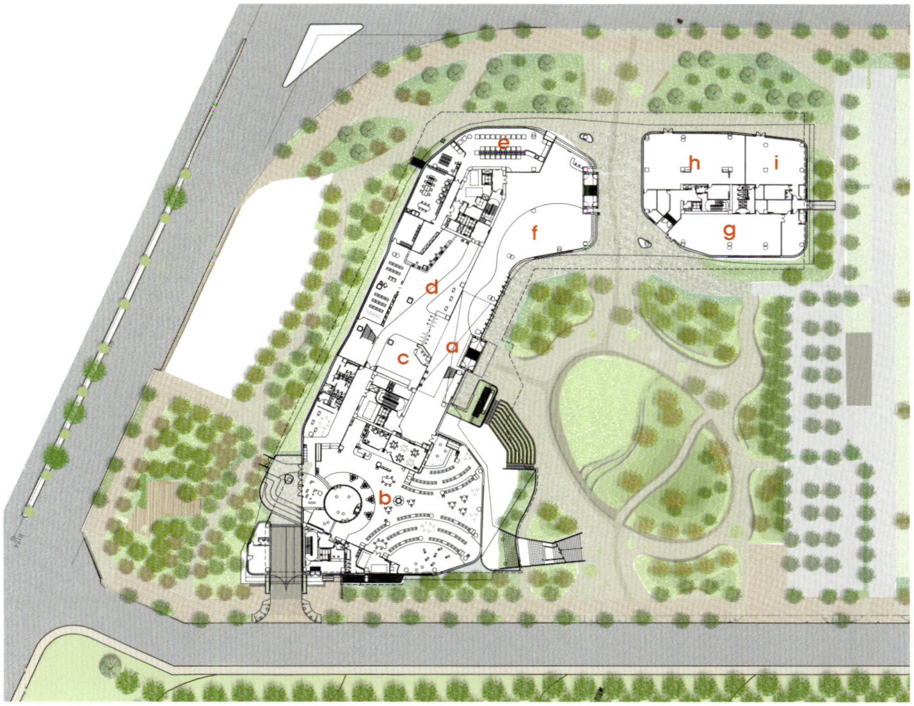

Ground floor plan 首层平面

a Lobby 大厅
b Children's center 儿童学习区
c Interactive directory 互动导览区
d Circulation 流动服务区
e Automated book return 自动还书区
f Digital experience 数字体验区
g Coffee shop 咖啡厅
h Bookstore 书店
i Convenience store 便利商店

1

2

3

1 Humanity and art area 人文艺术区
2 Science and technology area 科技产业区
3 Children's learning area 儿童学习区

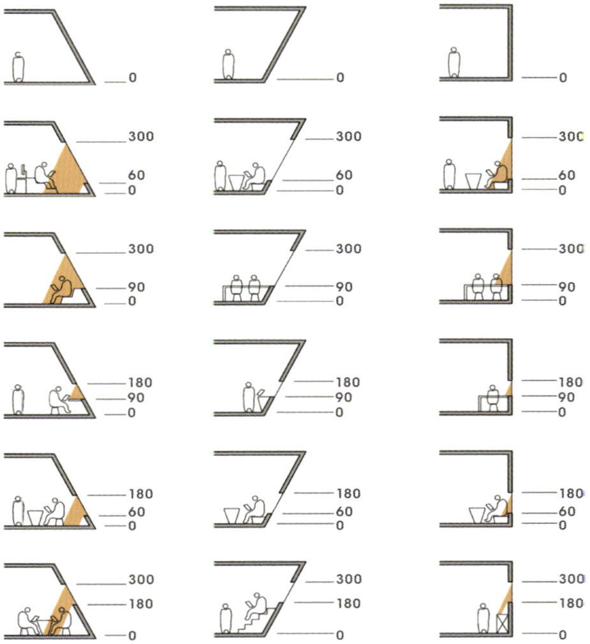

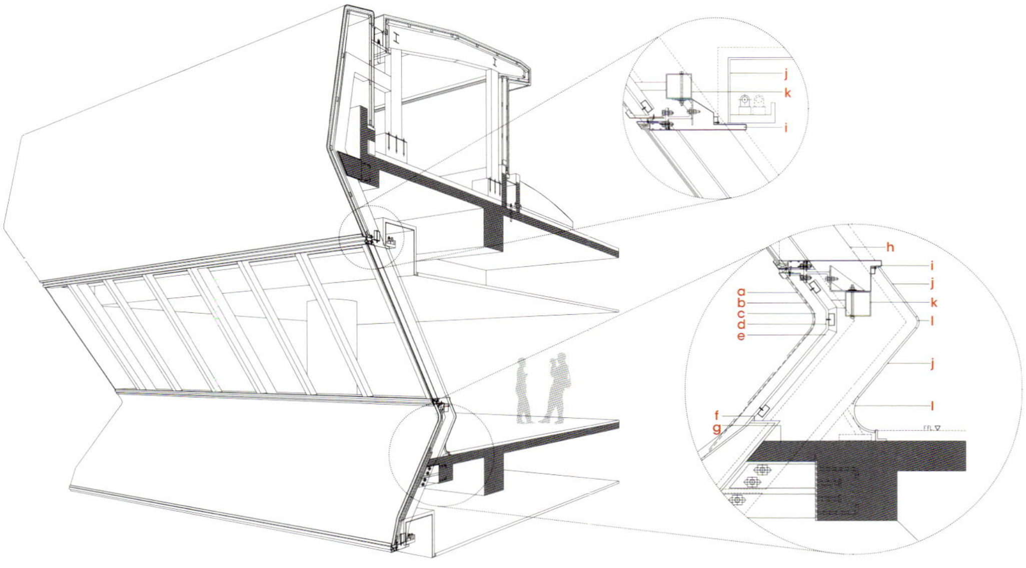

a Mosaic on adhesive base coat 瓷质马赛克
b Reinforcing mesh 抗裂玻璃纤维网
c Rigid insulation board EPS 板
d Sheathing 防水纤维板
e Steel framing 复合墙板固定铁件
f Firestopping wall 防火隔音墙
g Reinforce concrete structure 钢筋混凝土垫层
h Steel post 抗风柱
i Metal panel 铝板
j Drywall panel 内封纤维板
k Steel beam 抗风梁
l Glass fiber reinforced gypsum (GRG) 造型弧板

Wall section details 墙剖细部

Three Dimensional Curved Construction
多维度曲墙工法

The fluid building form and ever-changing spatial movement are achieved by a composite wall system including a structure system, exterior wall system, exterior wall decoration material, horizontal strip windows, and exterior wall extended structure. The main structure system consists of reinforced concrete, a steel structure system and tubular load-bearing walls. The exterior wall system is composed of lightweight steel frame and composite wallboard. To reflect the texture of ripples and pebbles, the folding surfaces are cladded with pearl-white round mosaic tiles of eight different sizes, enabling the skin to achieve a "monocoque" appearance.

多维度曲墙工法在主体结构系统、外墙构架系统、外墙装饰材料、水平带窗及外墙延伸构造的施工上展现重点。其中，主体结构系统以钢筋混凝土、钢结构系统、管状式承重墙所构成。外墙构架系统则以轻量化钢构架及复合式墙板组成；项目以新研发的“贴覆类材料”取代一般瓷砖，采用圆形马赛克作为主要贴覆材料，反映河川波纹与卵石之质感；其次，为避马赛克排列产生具有方向性之线条影响外观，经研发与测试后，选定八种不同粒径的马赛克组成一单位，每单位 (30 厘米 x30 厘米) 边界以锯齿状搭接，避免明显的接缝影响外墙自然波动流线。

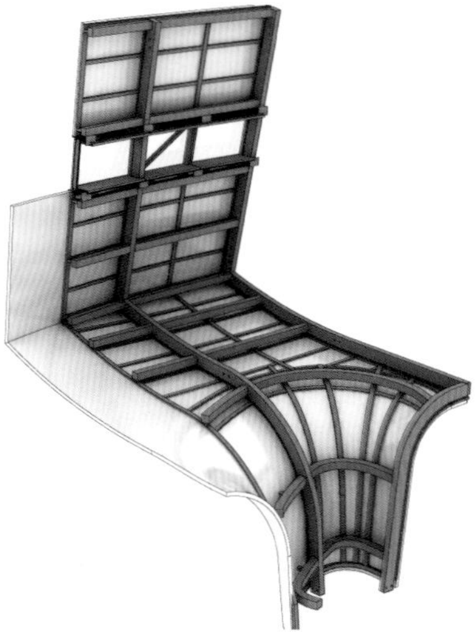

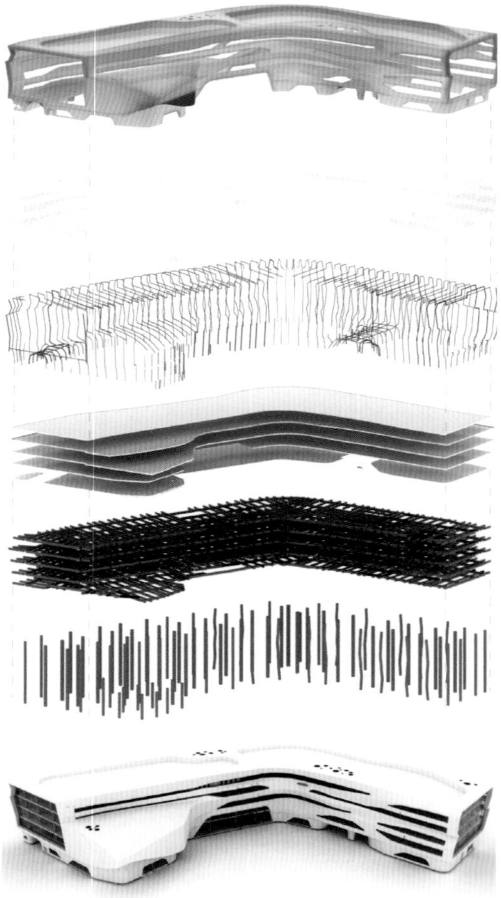

4

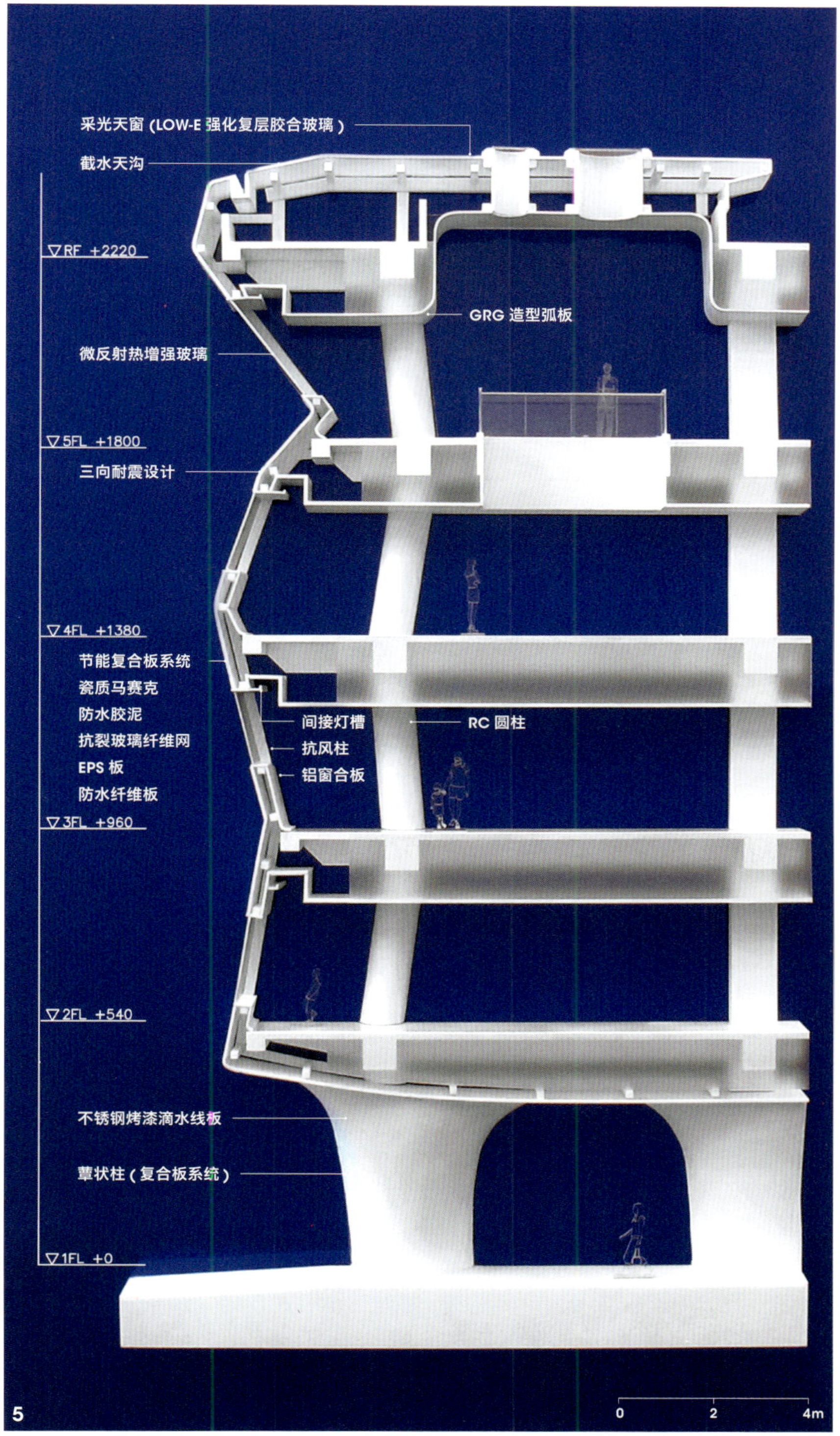

5

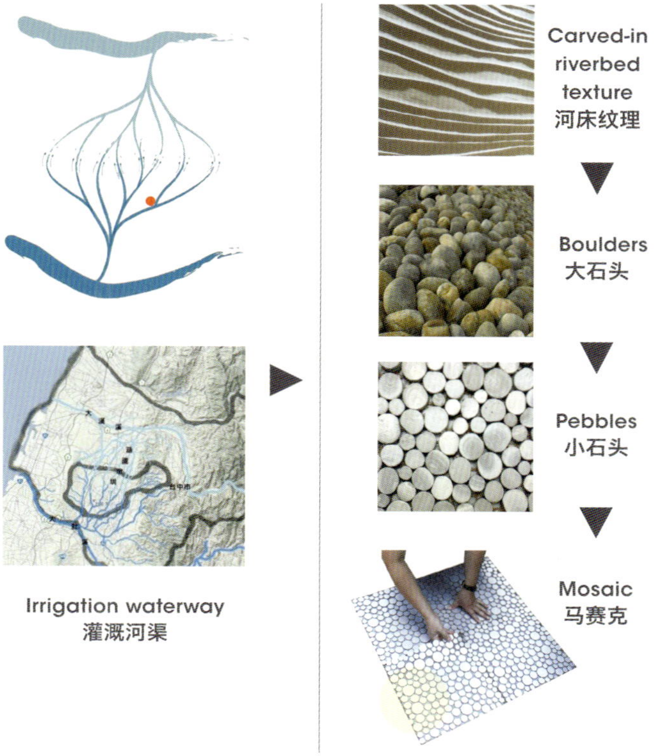

6

4 Structural analysis 结构分析
5 Section model 外墙剖面模型
6 Concept of exterior wall 外墙概念

Chocolate Museum
巧克力博物馆

2010–2012
Taoyuan
3,967 m²

Flowing Landscape of Taste
味觉的流动地景

By creating a sweet, joyful and romantic chocolate experience, this first chocolate museum in Southeast Asia was designed to promote chocolate as a specialty food. The building is conceived as a solid volume cut in different angles with a chocolate-colored exterior to convey the imagery of an exquisite piece of chocolate with a surprise filling. To achieve the desired richness and complexity of spatial experiences, the design utilizes a combination of RC and steel structural system, along with an exterior wall of aluminum and glass, to emphasize the contrast between the dynamic and static components. Such contrasts provide visitors with a vivid impression of tasting chocolate. A greenhouse of cacao trees is located in the center of the building to allow visitors to understand the origins of chocolate, thereby further appreciating the value of food resources and the environment.

项目为东南亚首座巧克力博物馆，借此推广巧克力欢乐浪漫的美食文化。在创意构思阶段，即以味觉出发，转化演绎出多种空间模型，最终以块体切削的方式，呈现“巧克力剖开流泄出内馅”的形态；从内部展览动线一路蔓延至室外成为流动的地景，如巧克力口感般滑顺而绵延不断。冲孔板与铝面板呈现于立面上表现为虚，钢筋混凝土与钢结构落于结构系统上是为实，空间上的虚实对比让访客于走读展览和品尝产品的过程中，体验到味觉所转化的空间层次感与张力；在绕行可可树温室的同时，了解巧克力的原料及生产流程，进而培养尊重食物资源与环境的核心价值。

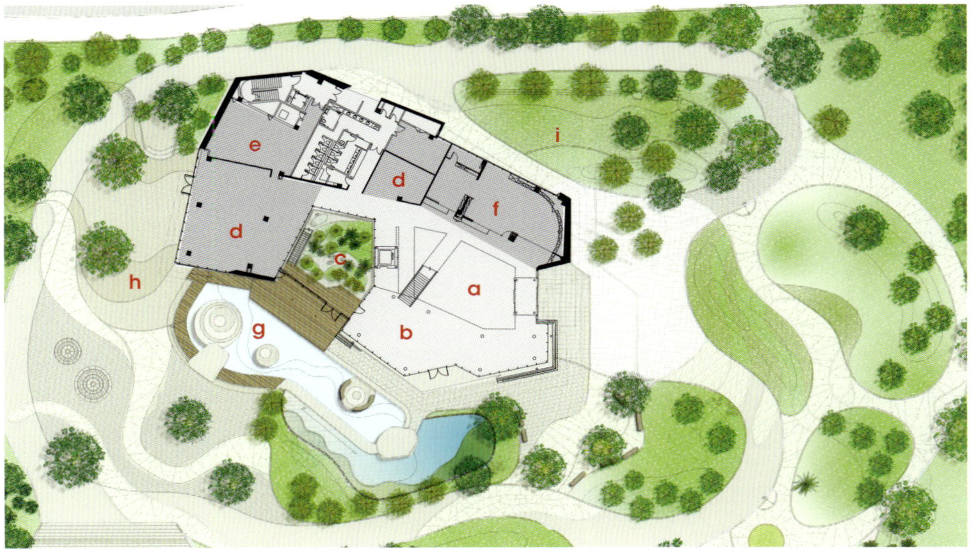

- a Lobby 大厅
- b Lounge 沙发区
- c Cacao greenhouse 可可树温室
- d Restaurant 餐厅
- e Kitchen 厨房
- f Gift shop 商品店
- g Interactive fountain 互动喷泉
- h Mist fountain 薄雾喷泉
- i Children's playground 儿童游戏区

Ground floor plan 首层平面

Good Design is Good Business
好设计引领商业新契机

Innovative space design inspires new opportunities for the transformation of traditional factory showrooms. The project successfully creates a tour by weaving together the dynamic and static activities in the museum and the solid and void spatial experiences. The non-grid configuration column structure further enhances the desired richness and complexity. As a building that caters to visitors, the Chocolate Museum is an exemplary model which facilitates the growth of local business and successfully transforms an industrial city into a recreational and fun environment for all.

通过创新的空间设计启发传统工厂转型，并借由外墙虚实呈现空间动态与静态之对比，以非矩阵柱梁、混合 RC 与钢结构以及虚实体验的活动空间，带来稳定的参观人潮，除增加地区的经济效益外，更进一步将制造业为主的工业城市空间转型为具有可亲性的休闲环境。

1

2

1 Lobby 大厅
2 Cacao greenhouse 可可树温室

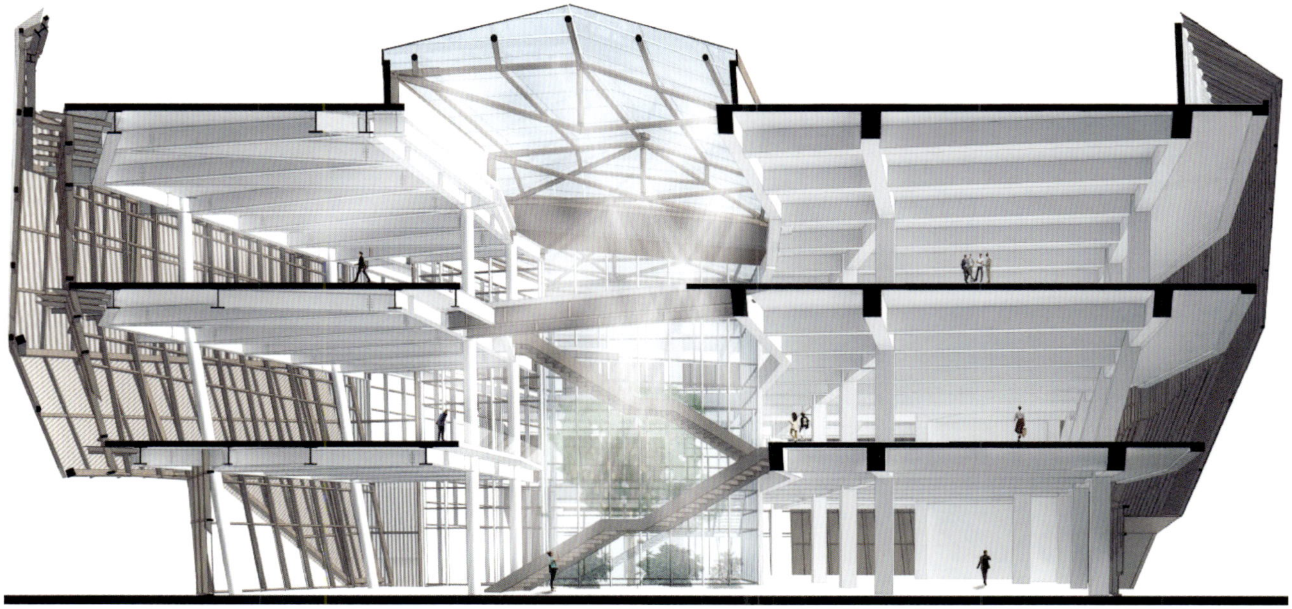

Perspective section 剖透视

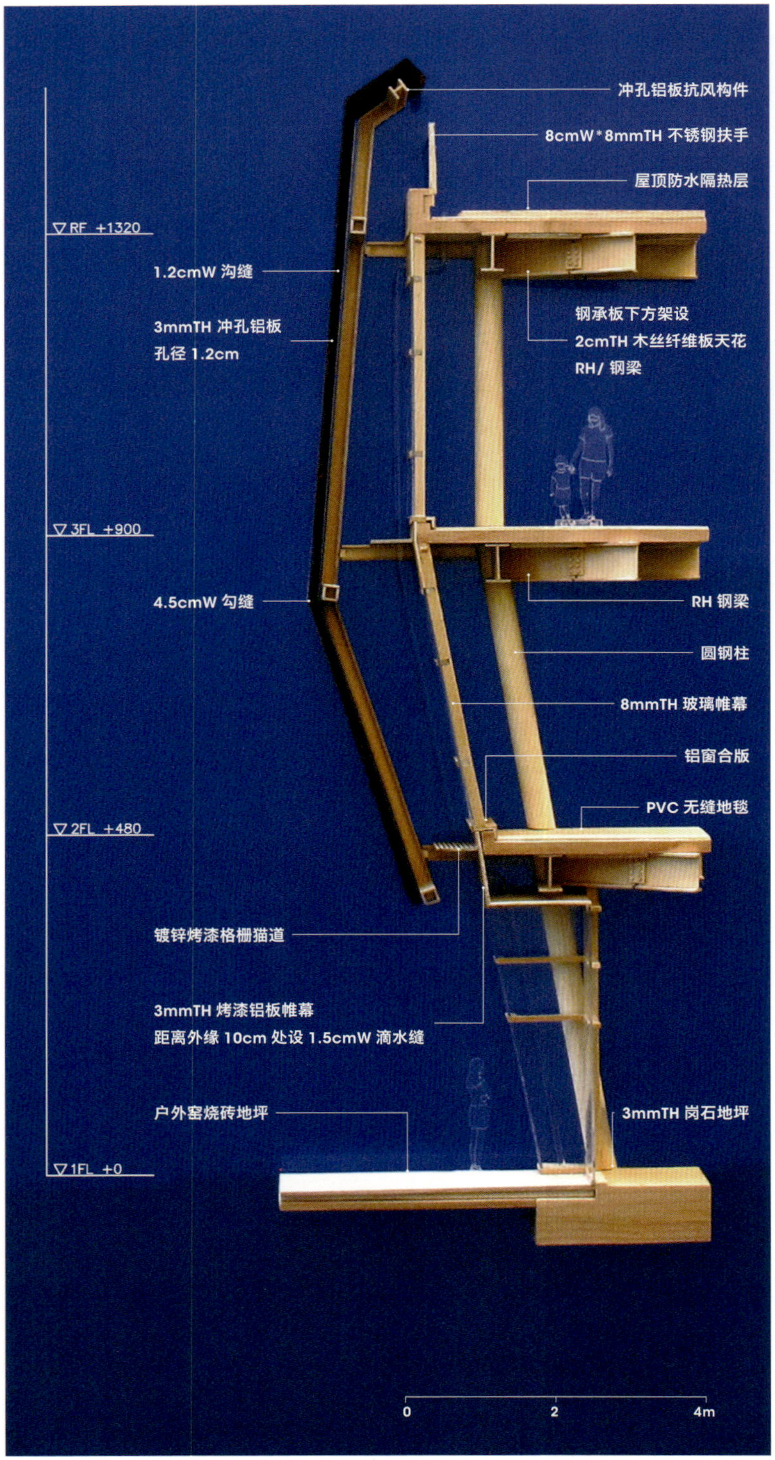
冲孔铝板抗风构件
8cmW*8mmTH 不锈钢扶手
屋顶防水隔热层
▽RF +1320
1.2cmW 沟缝
3mmTH 冲孔铝板
孔径 1.2cm
钢承板下方架设
2cmTH 木丝纤维板天花
RH/ 钢梁
▽3FL +900
4.5cmW 勾缝
RH 钢梁
圆钢柱
8mmTH 玻璃帷幕
铝窗合版
PVC 无缝地毯
▽2FL +480
镀锌烤漆格栅猫道
3mmTH 烤漆铝板帷幕
距离外缘 10cm 处设 1.5cmW 滴水缝
户外窑烧砖地坪
3mmTH 岗石地坪
▽1FL +0
0
2
4m

Advantech Smart Campus
研华智能园区

2011–
New Taipei City
36,600 m²

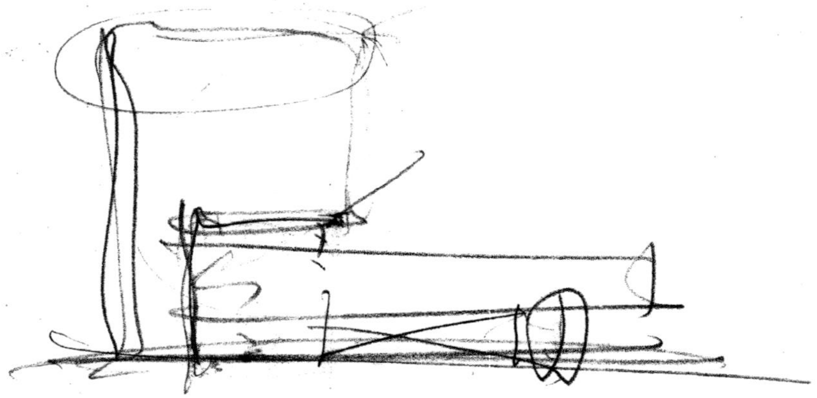

Optimal Campus Planning
最佳效益的园区配置

Covering an area of 3.5 hectares, the Advantech Smart Campus consists of four buildings constructed respectively in three phases; Phase I: Office Building and Intelligent Greenhouse; Phase II: Manufacturing and Distribution Center; and Phase III: Innovative R&D Center. The design intentions of creating a central axis, placement of a central courtyard, a distinct front from the rear are emphasized in an attempt to articulate a campus that, together with intelligent green building technologies, create a comfortable working environment in the Industry 4.0 era.

随着工业 4.0 时代兴起，JJP 团队结合智慧科技与绿色建筑，营造舒适便利的未来工作环境。全区占地约 3.5 公顷，共分三期四栋兴建，已陆续完成展示暨办公栋、智慧温室及主力生产厂房；并为创新研发中心预作规划。园区整体以轴脉（Axis）、段落（Phasing）、苑埕（Place）、向背（Front/Back）为设计准则，有效掌控分期拓建计划。

a Lobby 大厅
b Conference hall 国际会议厅
c Display room 智能展示室
d Restaurant 餐厅区域
e Finished product warehouse 成品仓储
f Plaza 星湖广场
g Office 办公室
h Smart hut 智能屋
i Central control room 中控室

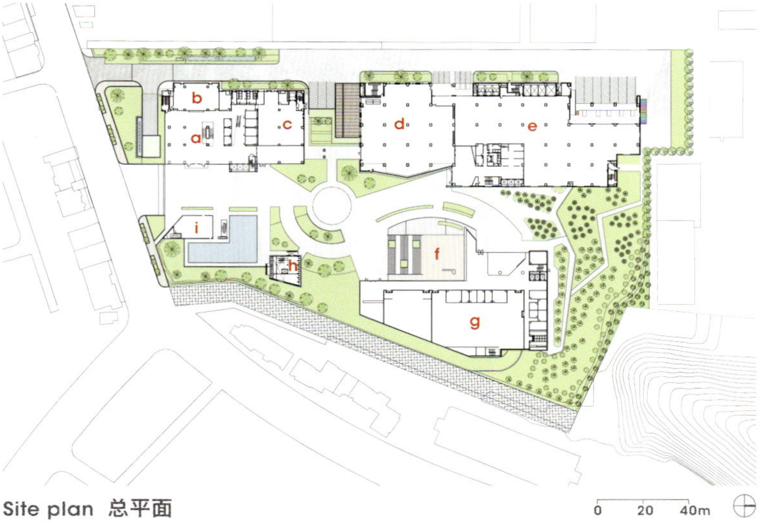

Site plan 总平面

Green × Intelligence
体现绿色建筑空间智能化

The building façade incorporates bricks and exposed concrete to create a unified campus style. These materials are chosen for not only their rich textures, but also their ability to reduce the heat load on the building exterior. Other design features such as brick cavity walls, deep recessed windows, insulated low-E glass, vertical shading panels, all enhance the performance of the building envelope. A comfortable and smart working environment is achieved by the integration of interior spaces and intelligent building technology. Examples include a lobby with a low-maintenance green walls, intelligent conference rooms and communal spaces, mobile offices and canteens, and a situation room that utilizes Industry 4.0 applications. Through the integration of the architectural design and intelligent building systems, the project gradually obtains the relevant certificates as it becomes a leader in the field of intelligent green architecture.

建筑群以清水砖及清水混凝土配搭营造校园风格，利用复层清水砖墙、深凹窗、Low-E 双层玻璃、垂直遮阳饰板灯柱及钢材等材料混搭，丰富立面表情并有效降低建筑物外壳热负荷。室内则透过建筑空间与智能装置整合，包括光合植生墙的挑空大厅、智能预约会议区、远程行动办公室、智能餐厅、工业 4.0 监控的战情会议室，以及厂房区两两挑空的主题交谊厅、开放式茶水间与随处可见的自助吧台等设计，打造科技互动实境，体现绿色建筑空间智能化的价值。园区透过建筑设计协调整合智能建筑系统接口与互动装置，融入创新的科技应用，并陆续取得多项标准认证，朝向未来全新绿色建筑智能园区领先地位迈进。

1

2

1 Lobby 大厅
2 Restaurant 餐厅

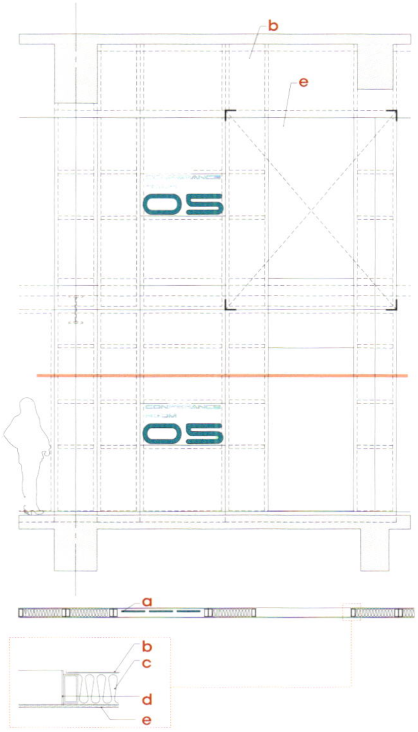

Glass wall 玻璃隔间

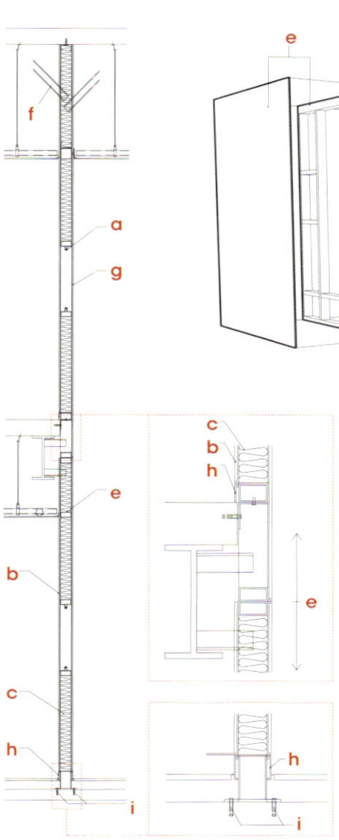

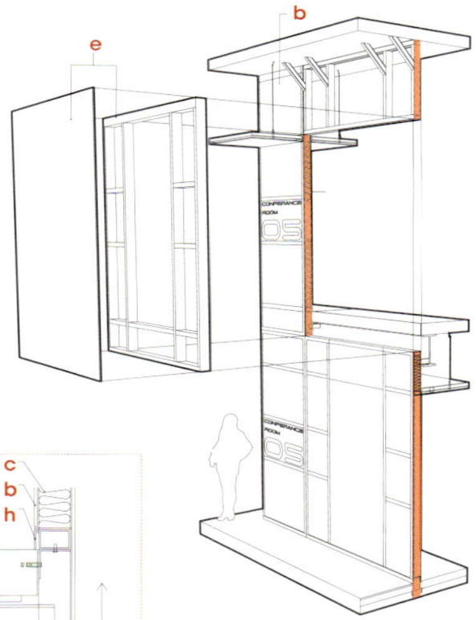

a LED light fixture LED 灯具
b High-density fiberboard 强化纤维板
c Soundproofing insulation 隔音棉
d Stainless steel panel (brushed finish) 不锈钢毛丝面
e Aluminum curtain wall system 氟碳烤漆铝挤形帷幕系统
f Galvanized reinforcing steel channel 镀锌补强槽钢
g Openings for intelligent room signage 烤漆玻璃透光区域
h Stainless steel baseboard 不锈钢踢脚
i Expansion anchor bolt 膨胀螺栓

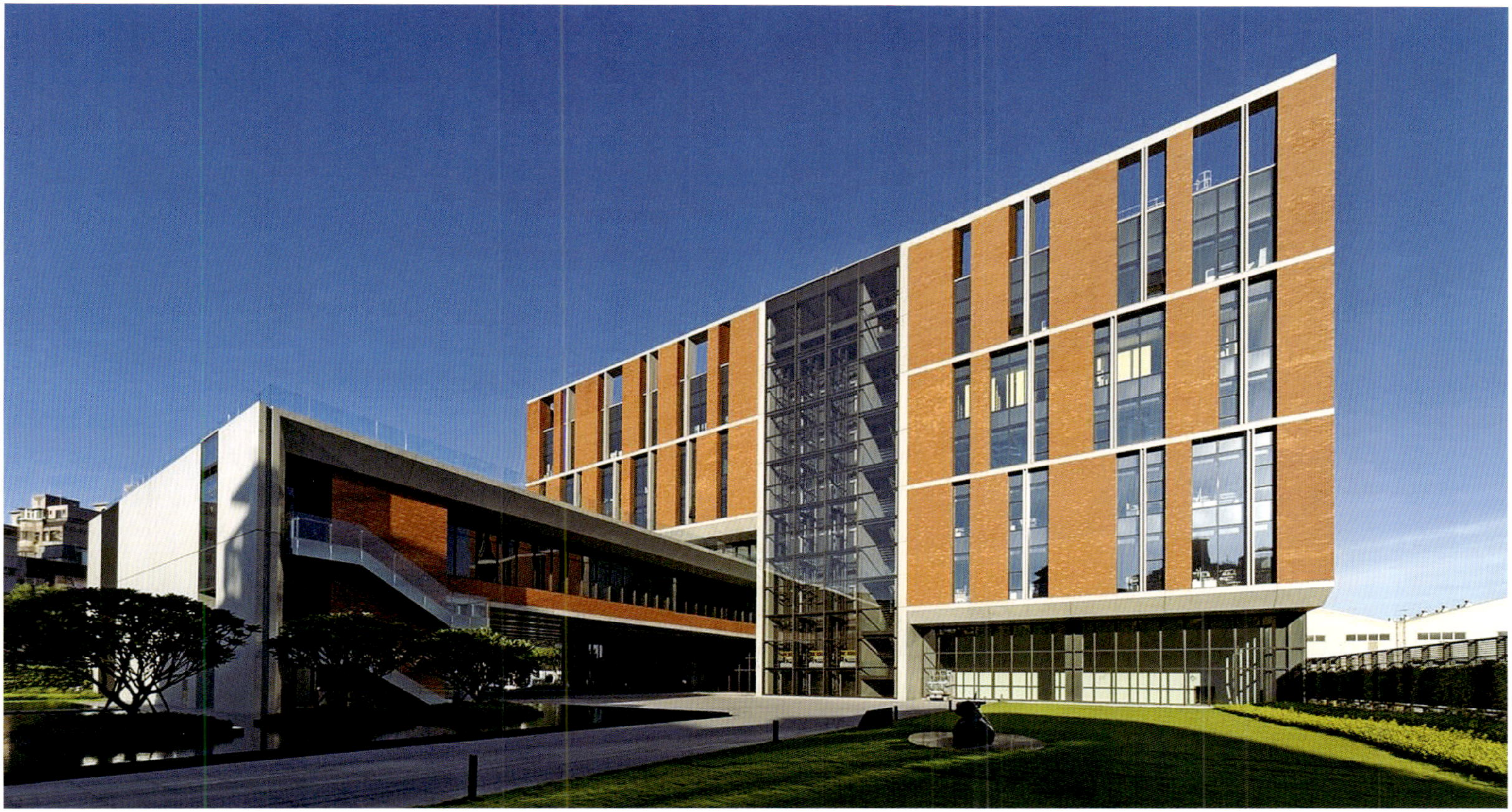

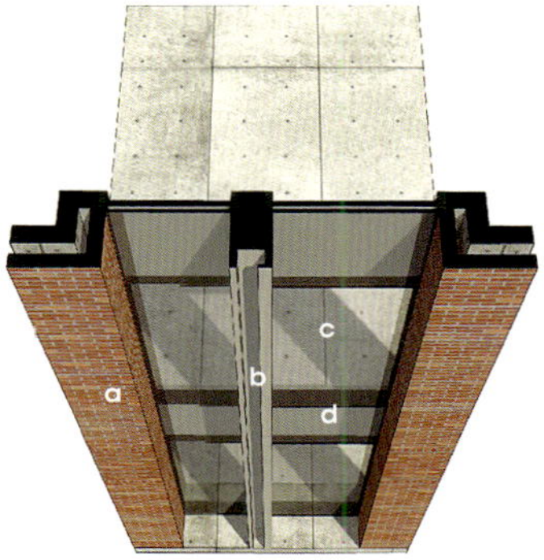

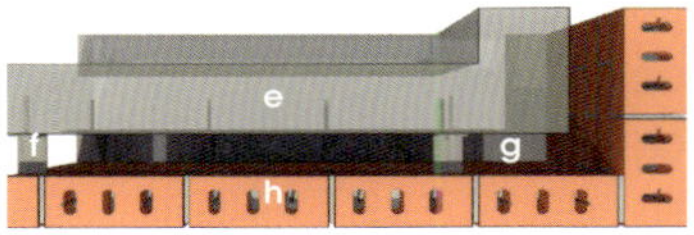

Cored brick wall 清水砖墙

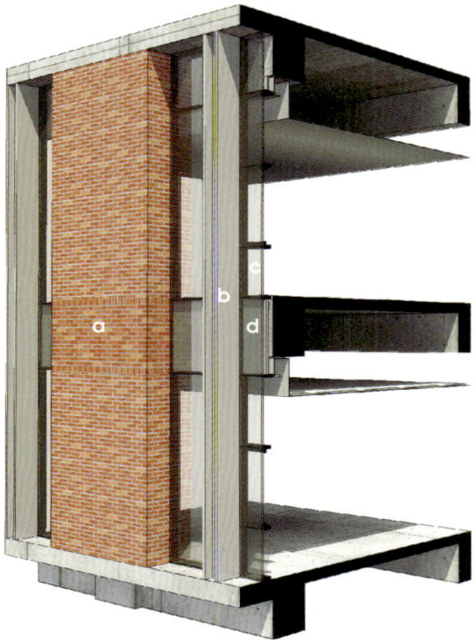

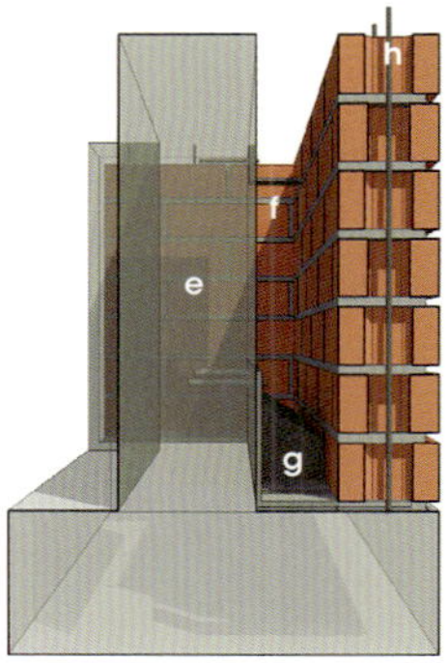

a Face brick 清水砖
b Metal fin with lighting 垂直铝板灯柱
c Low-E insulated glass Low-E 双层玻璃
d Spandrel glass Low-E 双层玻璃 + 背衬板
e RC structure 混凝土结构体
f Cavity ties 水平系件
g Horizontal metal anchor 水平承载铁件
h Vertical reinforcement 镀锌螺杆

NCCU Research and Incubation Center

政治大学研究暨创新育成总中心

2009–2014
Taipei
15,998 m²

----- Campus axis 校园轴线
▲ Campus gate 校园大门

Conversation Between Humanities and Science
人文与科技的对话

Situated at the end point of the existing NCCU campus axis, the project boasts a panoramic view that constitutes the visual focus and serves as a landmark in the area. The building mass is arranged in accordance with the variation in the topography of the site. The bright and dark contrasting colors of the two volumes reflect the basic spirit of the research center – interdisciplinary exchange and dialogue between humanities and technology. The lighter colored, more conspicuous, and dynamic mass symbolizes the free spirit of the humanities in academia, while the darker-colored, less conspicuous and more composed mass represents the precision and exactitude of technology. The two spirits converge and interact on the central tower, serving as a monument to knowledge in different research realms.

项目位于全校区制高点，拥有开阔全景的视野，因位于现有校园轴线的端点，自然而然成为地景上的视觉焦点。建筑物体量顺应地形高低错落配置，为响应本研究总中心的基本精神“人文与科技的对话”，外观以明暗两色材质对比区分，浅色体量代表自由、外显的“人文研究领域”，深色体量则为严谨、内敛的“科技研究领域”，互相交汇于中心标塔，开启跨领域研究的知识之窗。

Dual Courtyard Layout
双中庭合院式布置

In plan, the project adopts a double atrium courtyard configuration, introducing natural light and ventilation through staggered volumes. The two atriums give shape to a unique spatial interaction and link the upper and lower levels. Furthermore, a grand stepping indoor plaza serves as a gathering spot for researchers. The ground floor consists of a reception area, indoor and outdoor cafés, a landscaped terrace, exhibition room, lecture room, and a 200-seat auditorium. The second floor is mainly comprised of classrooms, offices, and research center workshops The entire third floor is dedicated to laboratories while the fourth and fifth floors are designated for research cubicles.

为创造研究人员停歇、交流的机会，平面规划采用双中庭合院式布置，引进自然采光通风，两处挑空的中庭串联上下楼层的休憩平台；一楼主要为咖啡厅兼接待区、户外咖啡座及观景平台、展示空间、阶梯讲堂及可容纳 200 人之国际会议厅；二楼主要为教室、办公室、研究工作室；三楼整层作为实验室使用；四、五楼为项目团队研究室及个人研究室。

Ground floor plan 首层平面

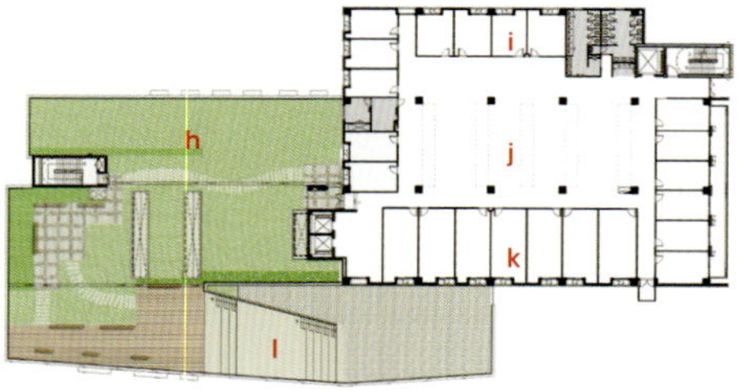

Fifth floor plan 五层平面

- a Hall 大厅
- b Lecture hall 国际会议厅
- c Exhibition 展示区
- d Classroom 教学教室
- e Film studio 数字虚拟摄影棚
- f Interactive classroom 交互式讲堂
- g Cafe/Restaurant 咖啡厅 / 简餐区
- h Roof garden 屋顶花园
- i Research room 个人研究室
- j Open-office 开放式办公室
- k Project room 专案研究室
- l Grand steps 景观阶梯

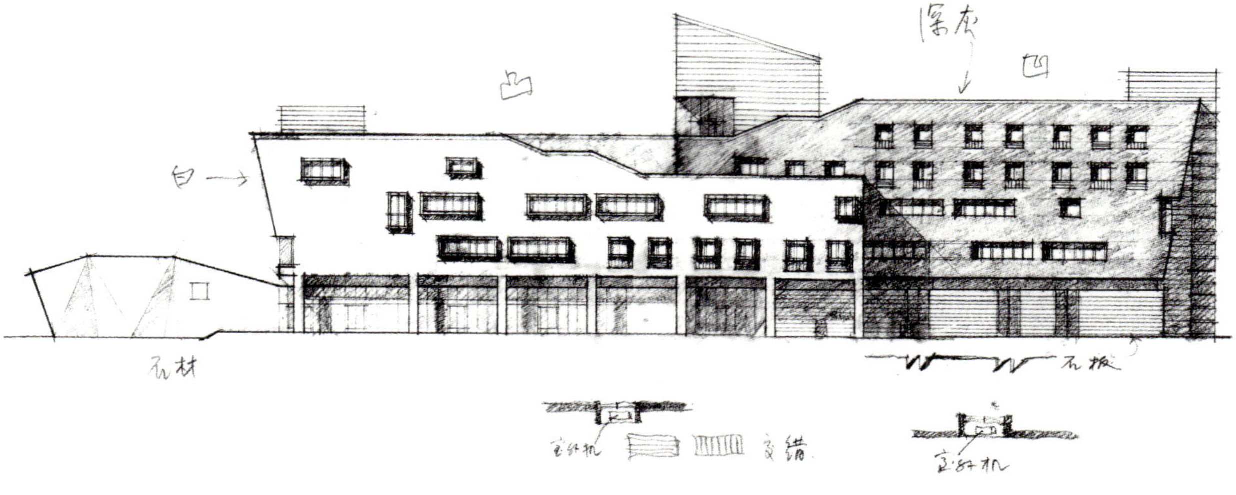
深灰
白
石材
石板

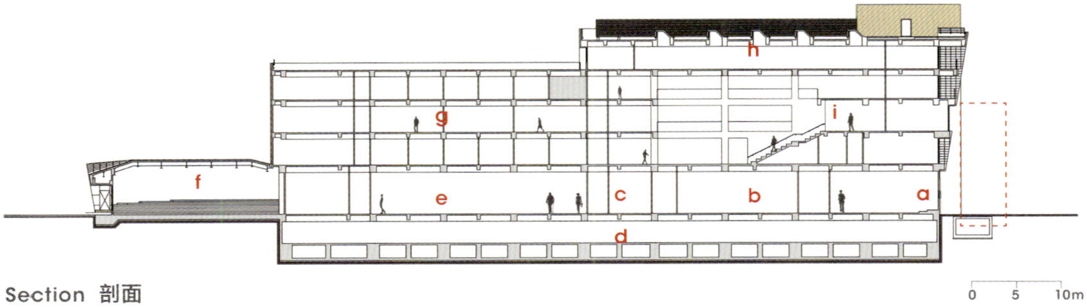

Section 剖面

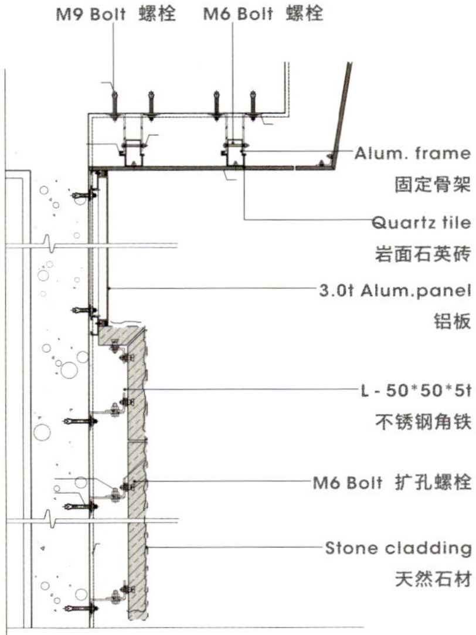

Wall section 墙体剖面

- a Interactive classroom 交互式讲堂
- b Film studio 数字虚拟摄影棚
- c Mechanical shaft 机械空间
- d Parking 停车空间
- e Exhibition 展示区
- f Lecture hall 国际会议厅
- g Laboratory 实验室
- h Office 办公室
- i Presentation atrium 成果发表空间

Taiwan Photon Source Facility
台湾光子源加速器实验中心

2007–2013
Hsinchu
54,835 m^2

Circular Accelerator
环形加速器及附属建筑

The three design guidelines for this project were to minimize impact on the landscape; to create a seamless integration between the new and the existing facilities; and to keep experimental facilities operational during the underground construction. The project is located on the west side of the Hsinchu Science and Industrial Park (HSIP), with the main building housing a circular accelerator with a diameter of 220 meters, which provides the world's brightest light source for the Synchrotron.

The exterior design uses "flowing light" as a primary theme, accompanied by a façade with varied divisions in a neat and lively manner. In order to preserve the forest in the northeast corner of the site, the Activity Center was placed in the southeast along with the existing buildings. Other auxiliary buildings are situated alongside the main road, which accommodate spaces for academic interchange and dining, 90 guest rooms and the MEP system required by the Taiwan Photon Source (TPS).

对环境负荷最小化、新旧建筑并存、维持高精密实验机能的正常运作是这个项目的三大设计指标。本工程坐落于新竹科学园区西侧，基地为 1/16 之缓坡地形，主建筑物为一栋 220 米直径的环形加速器建筑，用以装设全世界亮度最高的同步加速器光源。

外观设计以“光线流动”为主题，搭配层次感的立面分割，呈现简洁利落的风格；保留东北角之完整林地，将活动中心配置于东南侧现有建筑物之间，以维持原有空间纹理之完整；活动中心及机电三馆两栋附属建筑共构于区内主要道路旁，分别提供学术交流空间、餐厅及 90 间客房之住宿，同时供应加速器设备之特殊水电需求。

Five Major Challenges
五大工程挑战

Site plan 总平面

1. **Operation:** To keep two existing facilities operational during the underground construction.
2. **Installation:** A custom-made circular overhead crane was developed to install concrete slabs and laboratory equipment on site.
3. **Vibration:** To satisfy the vibration requirements: VC-F(1-80Hz, 1/3 Octave Band), which are the most stringent of any laboratory in Taiwan.
4. **Temperature:** The instrument control room and the protective shield surrounding the accelerator needed to be kept within a rigid range of 25±0.1°C.
5. **Radiation:** The design and construction of a protective shield surrounding the accelerator to prevent potential leakage of radiation materials. Two additional layers of concrete slabs separated by less than a 5mm gap were placed atop the protective shield for maximum safety.

1. **维持运作：** 现有小环馆与行政大楼于新建大环馆施工时，仍需维持正常运作并提供稳定的实验环境。
2. **施工吊装：** 应对水泥盖板吊装及设备搬运需求，研发出超大环型吊车，一并解决所有重型吊装问题。
3. **微振动抑制：** 符合台湾实验室中最高等级微振要求 VC-F (1-80Hz, 1/3 Octave Band)。
4. **温度控制：** 加速器屏蔽隧道区及仪控机房环境，温度控制须满足（25±0.1）℃标准，已趋近于恒温环境控制。
5. **防止辐射外泄：** 为使同步加速器有足够屏蔽防止辐射外泄，同时须兼顾设备吊装及更新作业，于屏蔽隧道上方施工两层厚度达 50 厘米的水泥盖板交叠封顶。此外，为确保辐射不外漏，所有水泥盖板间距均要求小于 5 厘米。

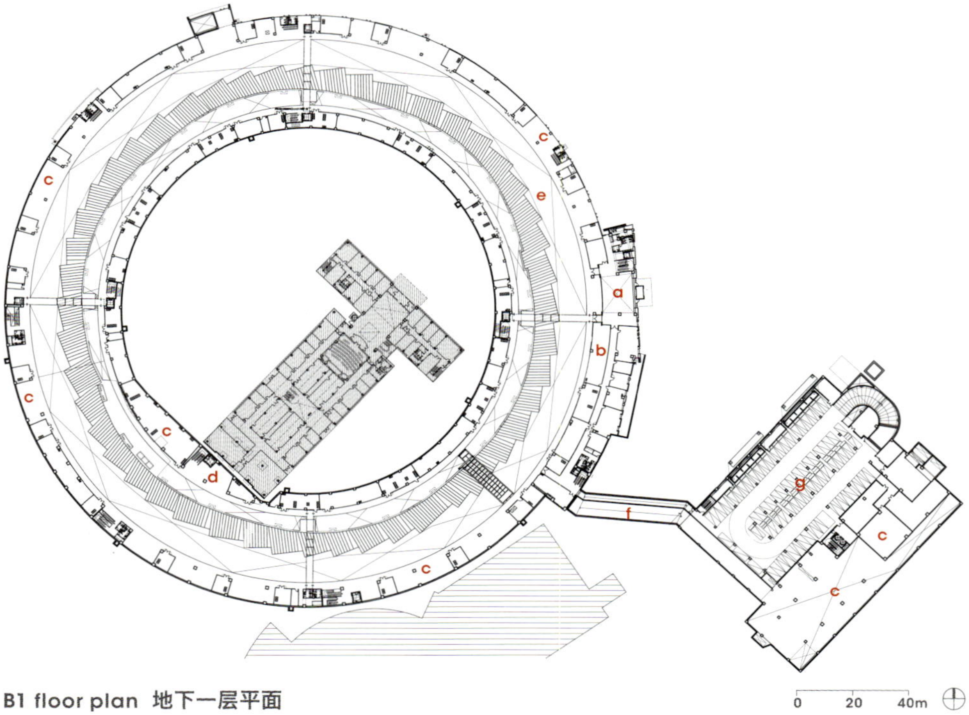

B1 floor plan 地下一层平面

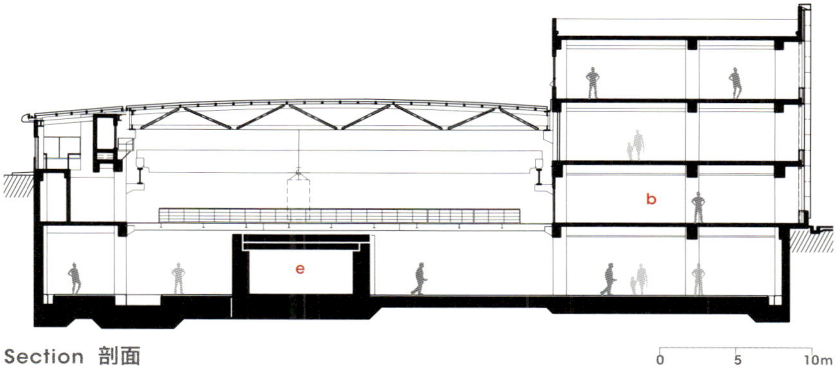

Section 剖面

a Hall 大厅
b Office 办公室
c Mechanical room 机房
d Stair hallway 梯厅
e Accelerator 加速环
f Passageway 连通管道
g Parking 停车场

Neo Solar Power Headquarters
新日光能源总部大楼

2007–2009
Hsinchu
47,940 m^2

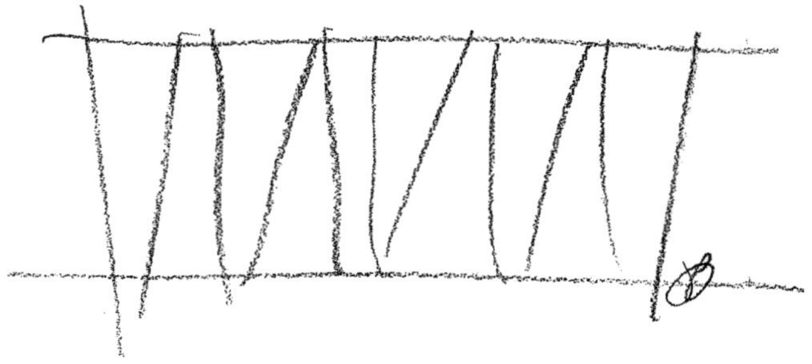

The Craft of Wave Effect
传递波动意象的建筑工艺

Facing a major access road coming off of the main boulevard of the Hsinchu Science Park, NSP's building complex borders an 80 meter-wide greenbelt. The design strategy was to induce greenery indoors and to present a precise, light and transparent look for the building, while respecting the client's desire for a "castle-like" image. The design concept emphasizes the links between NSP's corporate identity and its advanced photovoltaic production technology. The photoelectric effect, which is the basis of solar energy, is abstracted in architectural form by transforming the shape of the sine-curved light wave into a faceted glass curtain wall. The zigzagging floor slabs holding the wall panels in place follow the shape of the building. Eight distinct specifications of trapezoid units, with 350 different sizes of glass panes, were designed to achieve the complex 49.5° inclining and 96.7° sloping angles, optimizing construction and maintenance of the complex curtain wall.

项目位于新竹科学园区重要街廓转角处，东北侧紧邻 80 米中央绿化带。考虑业种之特殊性，以现代工艺技法诠释如城堡般的总部大楼意象，在整体空间营造的策略上，将外围绿化带景观延引入封闭的工作空间内。为表达太阳能转换电能的形式，外墙造型以半单元式帷幕系统隐喻连续波动的概念，帷幕折面分为 8 种梯形单元，分割尺寸约 350 种，各梯形单元外墙面分别前倾 49.5° 及后仰 96.7°，将各层楼板外缘皆依外墙进退面变化，收边钢梁随楼板弯折，作为幕墙之固定端，严格管控幕墙系统施工误差，展现建筑工艺技法。

a Lobby 大厅
b Conference room 会议室
c Canteen 餐厅
d Staff entrance 员工出入口
e Warehouse 仓库
f Fab 厂房
g Auditorium 礼堂
h Mechanical room 机械室

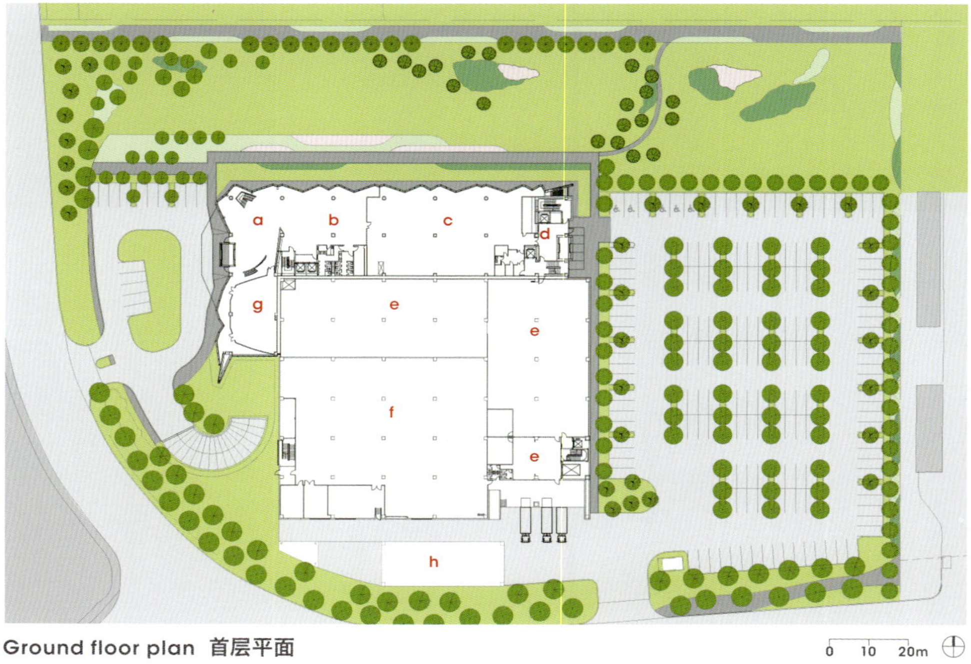

Ground floor plan 首层平面

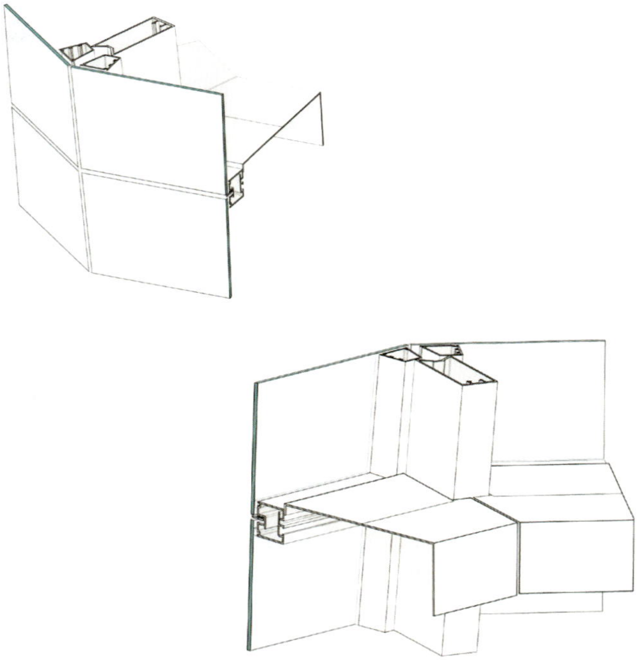

Curtain wall study 幕墙细部研究

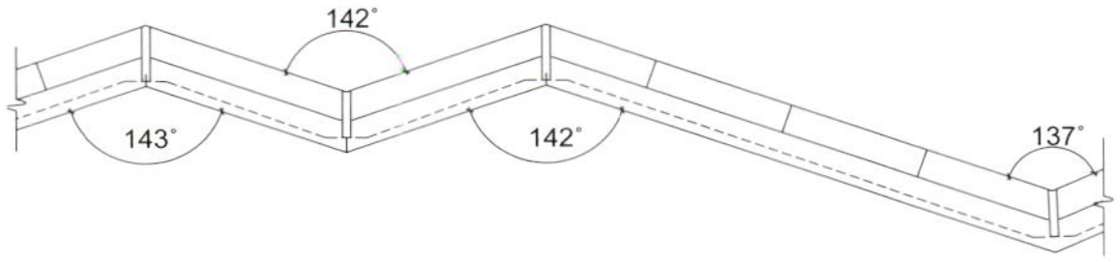

Wall profile type A 幕墙样式 A

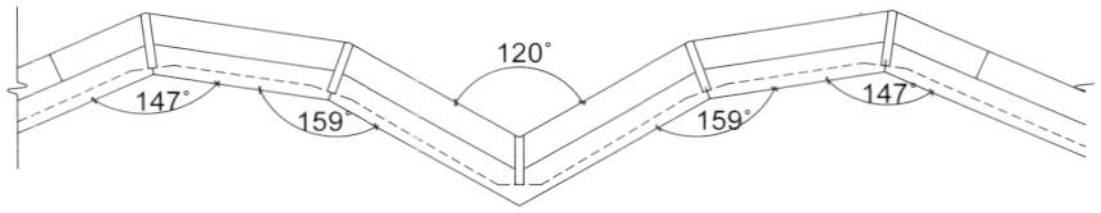

Wall profile type B 幕墙样式 B

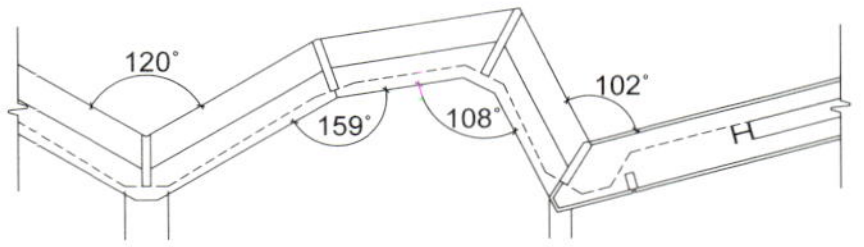

Wall profile type C 幕墙样式 C

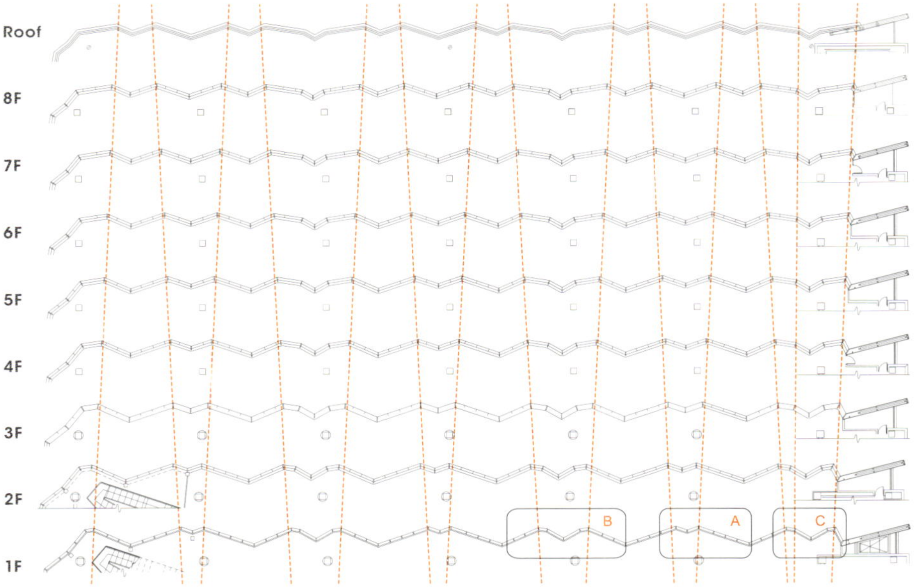

North exterior wall profile 北侧幕墙样式

Ally Logistics Yangmei Center
永联杨梅物流中心

2015–2018
Taoyuan
100,507 m²

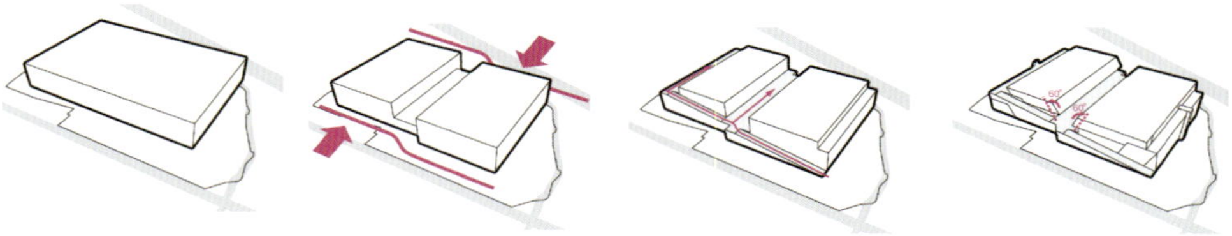

Innovative Logistical Architecture
创新物流建筑

To provide a highly efficient truck line and a brand-new warehouse form, the project aims to create a new benchmark for the next generation of logistics buildings, moving from traditional warehouses to modern logistics infrastructure with an international perspective. The three-tier storage of the "three-dimensional stack" is one of the characteristics of the project. In order to maximize unloading space, unloading areas are set on both sides of the ground floor; while the second and third floors are readily accessible from the ground level by truck. To prevent the building from appearing oppressive, the exterior design adopts 60-degree angle cement curtain wall, that integrates the interior wind bracing columns, balconies, truck parking areas, exhaust ports, air conditioning and structural systems, all the while creating a dynamic visual effect. With an eye toward the logistics architecture of the future, the external walls consist of pre-cast corrugated hollow cement panels, which have considerable advantages such as varied texture, heat insulation, and small displacement value.

项目目标为创造新世代物流建筑新标杆，从传统仓库迈向与国际同步的现代化物流基础设施，整合设计端与施工端的沟通流程，以提供高效率的货车动线，提出全新的仓储形态。“立体堆栈”的三层仓储是设计的特色之一，为满足三层空间卸货区最大值，卸货区除了设置于第一层楼两侧，二、三层皆以简明的货车坡道通达地面。外观采用 60° 斜角水泥板帷幕外墙设计，整合室内侧外墙抗风梁柱、阳台、货车停车位、排烟口、空调设备与结构系统，同时又创造出动态的视觉效果，以减轻体量的压迫感；外墙大量采用预制中空水泥板，具备质感多样，以及隔热、变位值小等优点，创造全新的未来物流建筑。

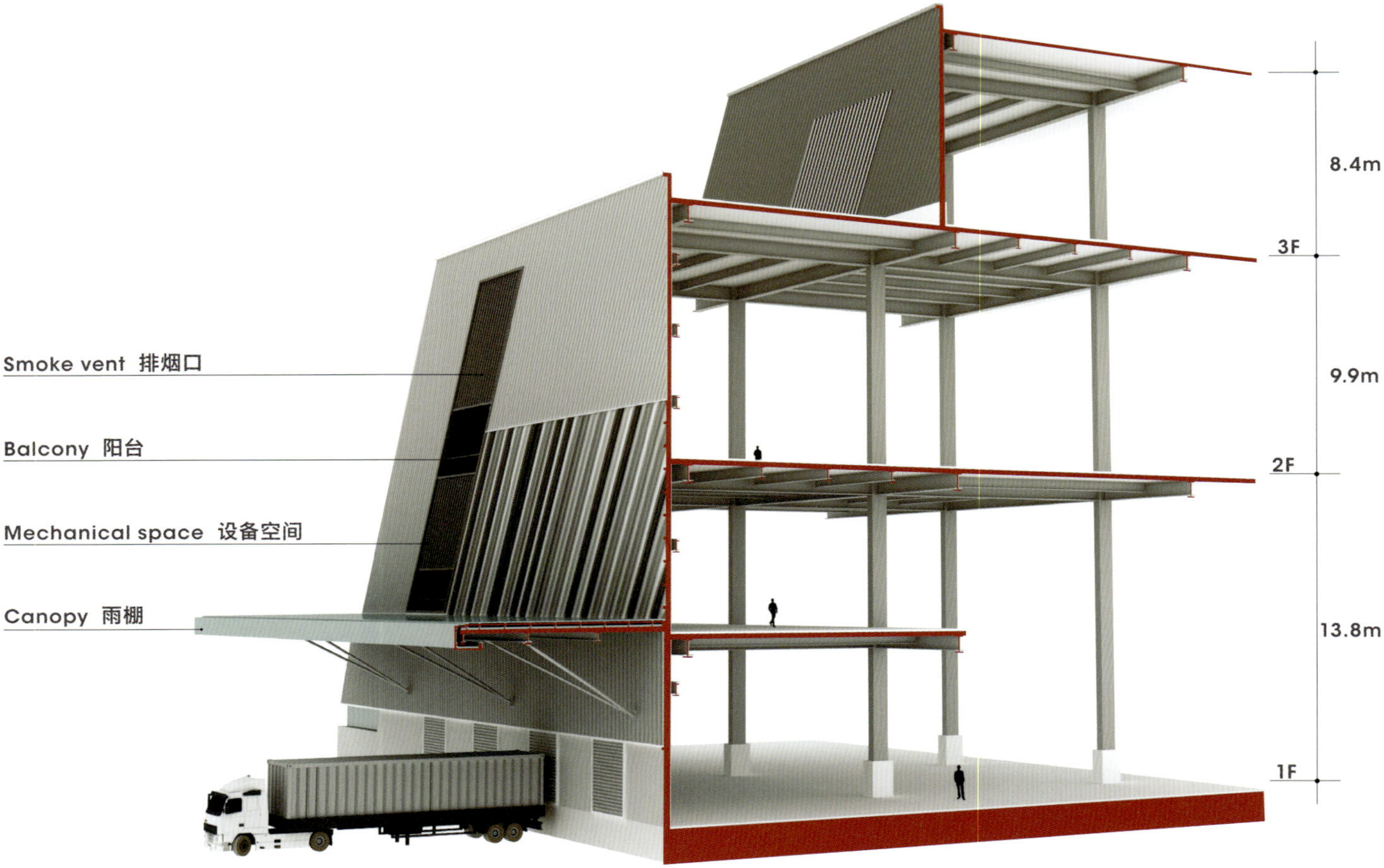

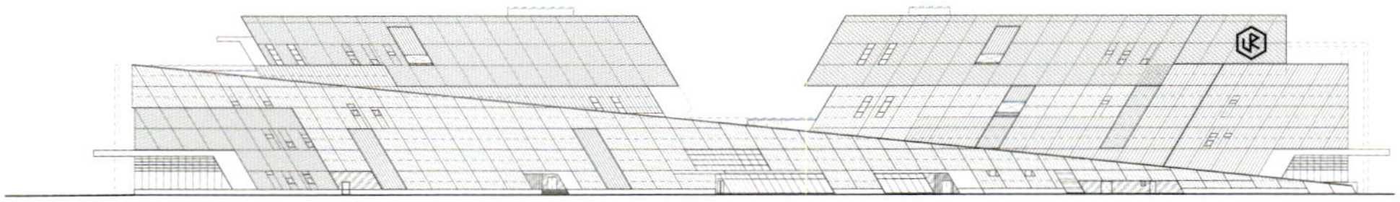

South elevation 南向立面

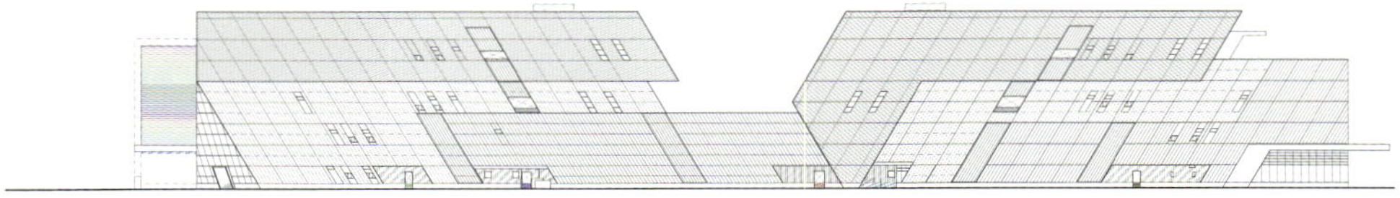

North elevation 北向立面

0 15 30m

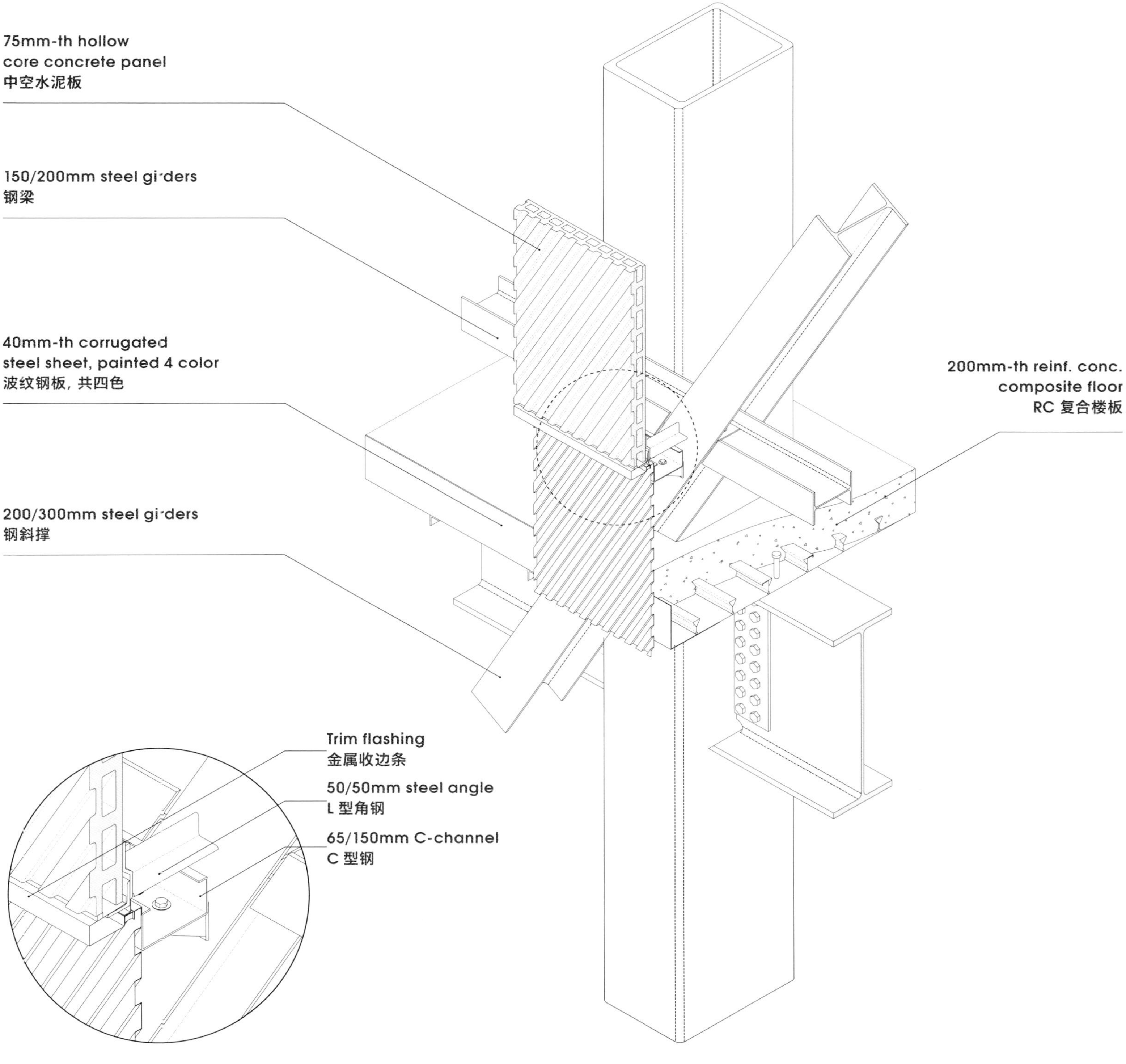
75mm-th hollow
core concrete panel
中空水泥板
150/200mm steel girders
钢梁
40mm-th corrugated
steel sheet, painted 4 color
波纹钢板，共四色
200/300mm steel girders
钢斜撑
200mm-th reinf. conc.
composite floor
RC 复合楼板
Trim flashing
金属收边条
50/50mm steel angle
L 型角钢
65/150mm C-channel
C 型钢

Giant Bicycles Headquarters
巨大机械总部大楼

2015–
Taichung
33,000 m^2

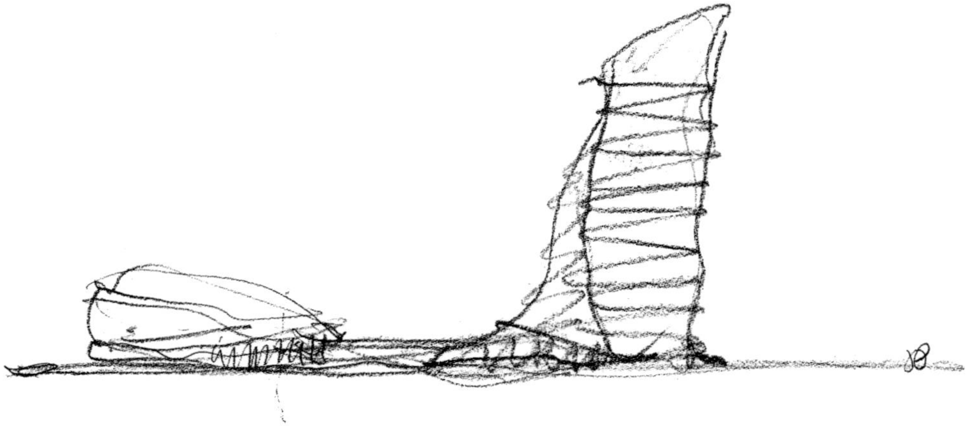

Fluid Tower
流动的塔

Located within the Central Taiwan Science Park, the Giant Headquarters project will undoubtedly become a landmark upon completion. In terms of architectural design, the building form is dynamically sculpted to evoke Giant's technological prowess, while its curvaceous shape and sustainable design features evoke the simplicity and eco-friendliness of the bicycle. The tower design takes full advantage of its unobstructed surroundings with floor-to-ceiling glazing on three sides and a generous 4.5m floor height to allow in ample natural light. Appropriate shading is provided by the architecturally expressive undulating slab covers and deep south-facing balconies. The expressive form of the exhibition hall along with an open plaza completes the overall composition.

项目位于台中科学园区，为了传递骑乘自行车的速度感，以“巨大”的企业品牌作为空间雕塑的设计原型，发展出两栋流线型建筑——办公栋及展示栋。办公栋的楼层透过梁带深度之渐变造型，除了提供适当的遮阳，也诠释“巨大”求新求变的企业精神；展示栋为自行车文化馆，流畅的动线系统由下而上形塑出整体的建筑和景观。骑行者可自由穿梭于两栋之间的活动广场，并沿着户外连续坡道衔接至各入口，再延续至周边景观环境，期待创造一个全新的自行车体验之旅。

a Office 办公楼
b Exhibition 展示空间
c Plaza 广场

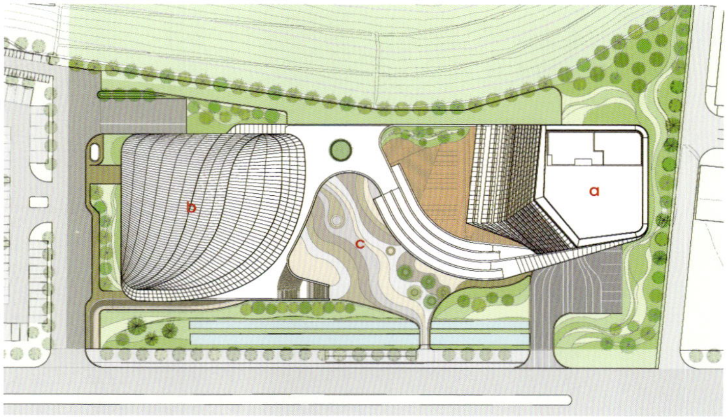

Site plan 总平面

a Lobby 大厅
b Parking 停车场
c Semi-outdoor plaza 半户外广场
d Exhibition space 展示空间
e Multi-purpose space 多功能空间
f Office 办公室
g R&D 研发中心
h Auditorium 会议厅
i Connecting bridge 连桥
j Cafeteria 自助餐厅
k Roof garden 屋顶花园
l Mechanical room 机械室

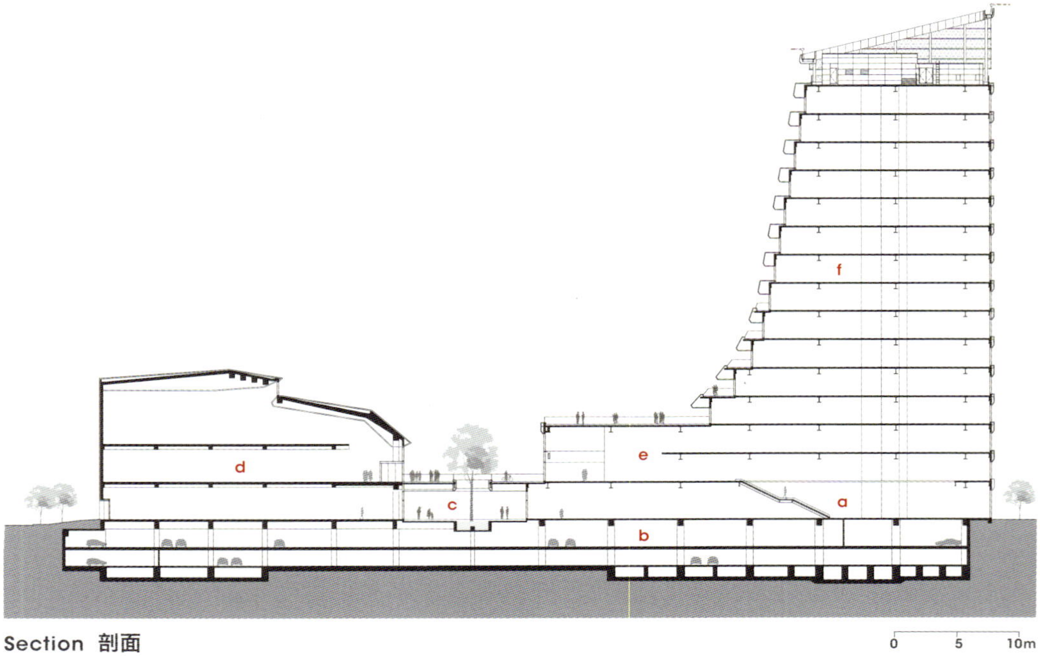

Section 剖面

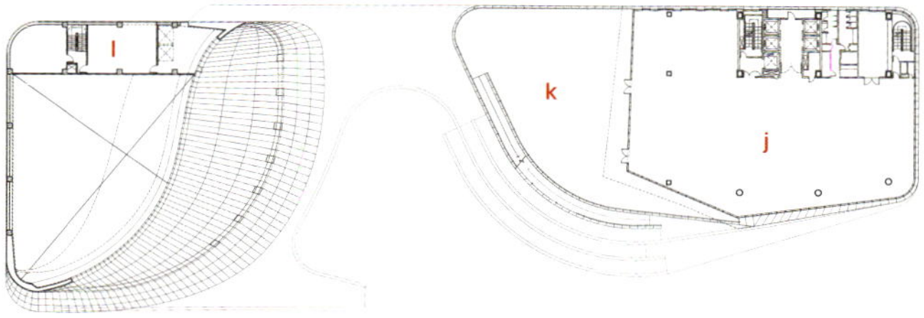

Fourth floor plan 四层平面

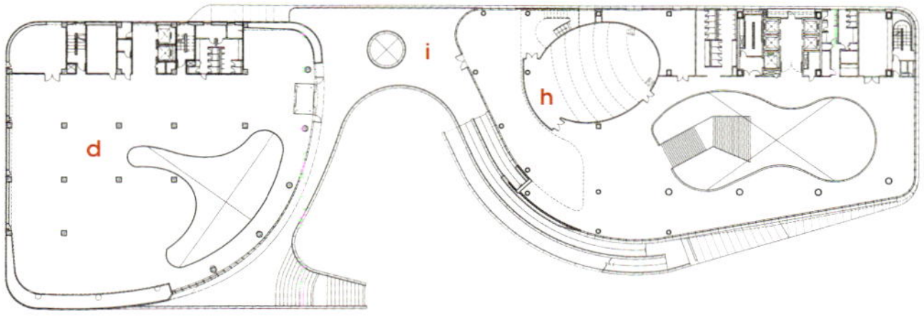

Second floor plan 二层平面

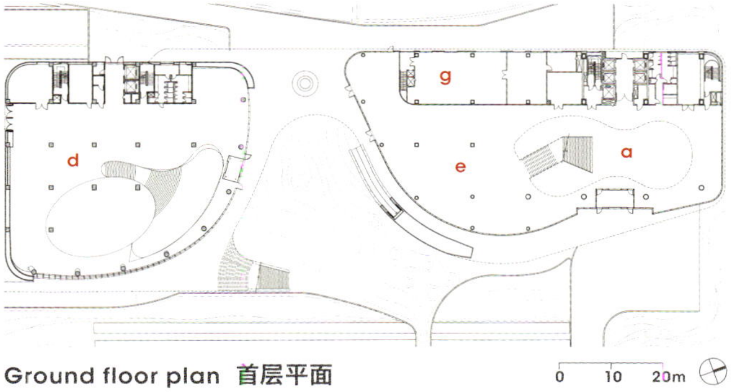

Ground floor plan 首层平面

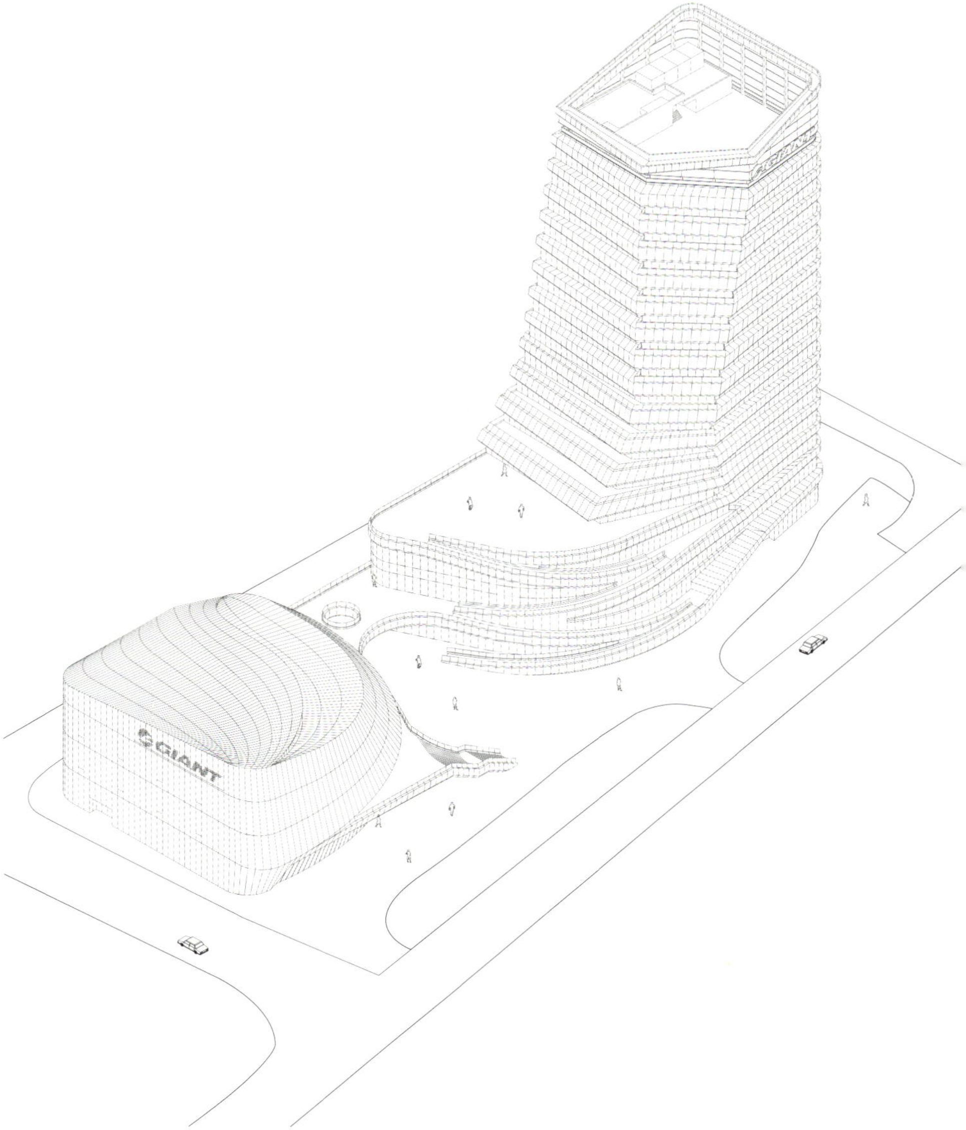

NCCU Center for Public and Business Administration Education

政治大学公共行政及企业管理教育中心

2016–
Taipei
27,450 m²

1

International Village of a Vibrant Community
小区共享的国际聚落

Integrating continuing education, international research and cultural life, the project builds upon four concepts: cultural innovation, social gathering, urban environment, community interaction to foster social, academic, and professional exchange. Through the analysis of spatial characteristics, function and configuration, two solid masses containing the "exhibition area" and the "residential area" are linked by a series of transparent volumes. The design delineates easy, navigable and flexible circulation. All corridors and vertical circulations converge at the 30-meter high central atrium. Glass cubes dramatically protrude into this central atrium, enhancing the overall spatial experience while subtly generating visual and symbolic connections among different users. In addition, by retaining a large number of mature trees on site, and planting multiple layers of native vegetation, the project will create an urban oasis setting for an international village of educational institutions and a vibrant cultural community.

本改建项目提出四大设计策略——创意凝聚、人文荟萃、生态节点、小区共享，整合进修教育、国际研讨及人文生活，成为产学交流的平台。建筑分为两大区块——会展区与住宿区，通过对空间特性、机能及配置的分析，创造兼具效率且饶富趣味的建筑，并以通透的体量串起公共及商业活动；挑高 30 米的流光中庭汇集所有会展动线，交错附挂于挑空面上的一个个玻璃盒研讨室与大厅之间产生张力。此外，项目延续城市的绿色生活，保留地面上大量的既有老树，并新植原生乔木及复层植栽，以打造出教育机构、人文小区与生态绿岛共构的国际聚落。

Cultural innovation
创意凝聚

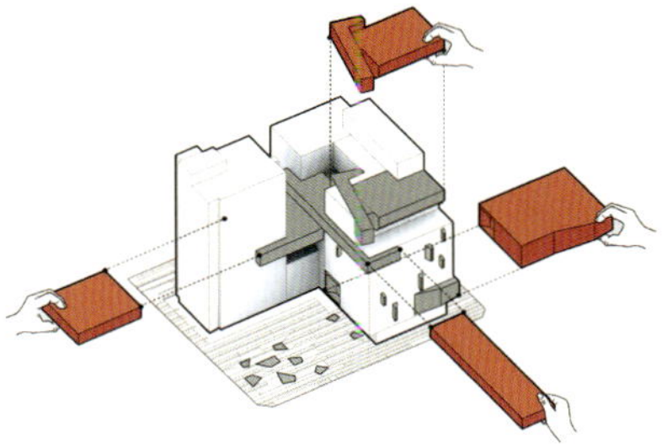

Social gathering
人文荟萃

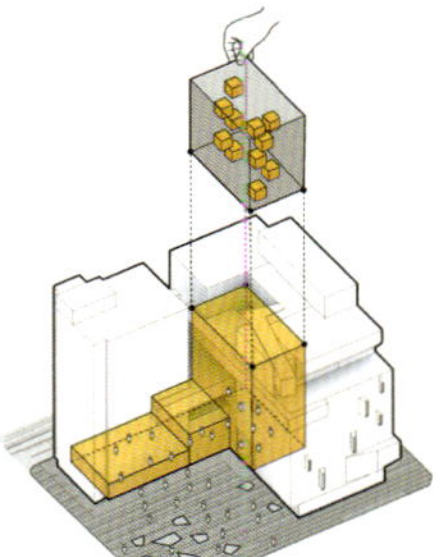

Urban environment
生态节点

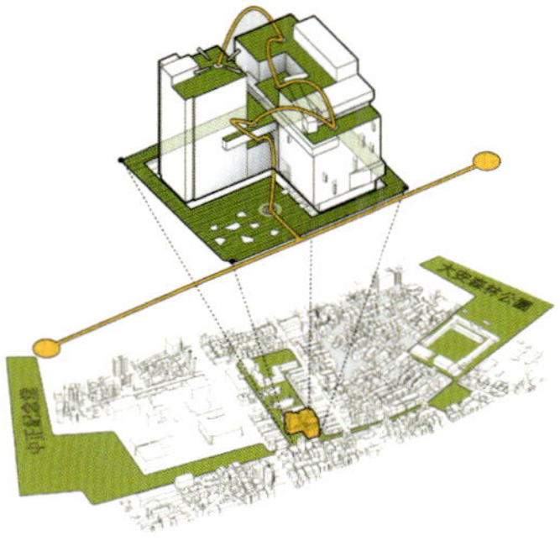

Community interaction
小区共享

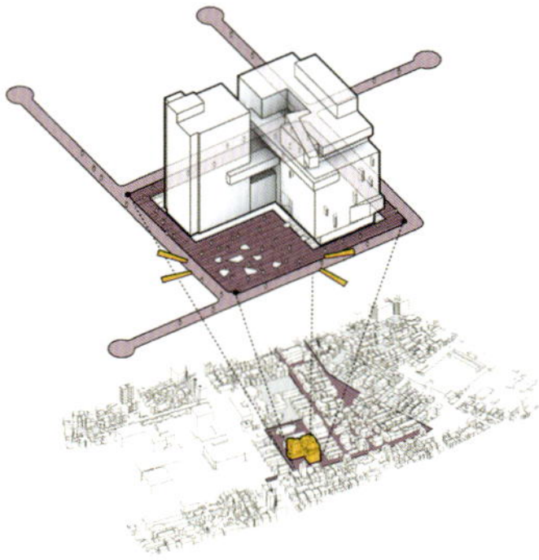

1 View from neighboring residential alley
自邻里巷弄端观景
2 Glass meeting box in atrium
玻璃会议室

1

2

03 SOCIAL

Address changes in society through constant compassion, research and experimentation in architecture

面对多元转变与快速更迭的时代，透过关怀、研究、实证连接社会空间与建筑形态

社会影响

IMPACT

DIA-LOGUE

CUI KAI x JOSHUA JIH PAN

Cross-strait Architectural Practice Environment and Social Impacts

崔愷 × 潘冀

两岸建筑执业环境差异与社会影响

Cui Kai 崔愷

Deputy President and Chief Architect of China Architecture Design & Research Group
Academician of China Engineering Academy
中国建筑设计研究院副院长、总建筑师
中国工程院院士

Joshua Jih Pan 潘冀

Founding Principal of J.J. Pan and Partners
Fellow, American Institute of Architects
潘冀联合建筑师事务所创所主持人
美国建筑师协会（AIA）院士

Time/Location: Winter, 2016, Land-Based Rationalism D·R·C, Beijing
对谈时间 / 地点：2016 年冬，北京本土设计研究中心

The Artisan Spirit — a Trend Toward the Development of The Specialized Talent

CUI: I would like to know more about vocational education in Taiwan. I still remember when I first joined the Design Institute after graduation, there were two types of technical personnel. The first type was architects and engineers; the second type was senior draftsmen and junior draftsmen or technicians, who mainly came from the secondary vocational education system. Later on vocational education became less popular in mainland China. While there are still employees from vocational secondary schools working in construction companies, such is not the case in the architectural industry. In the large design institutes today, most employees are graduate students. The newcomers assist the senior designers with the drawings but they usually aim to become design directors or even 'starchitects'. Therefore, design institutes are lacking professional architects who can make very precise drawings. I think this is one of the reasons why the level of architectural design and construction in mainland China is less than ideal. We simply do not have a stable design team with artisan spirit. It is a big problem.

Therefore, D·R·C welcomes people who are good at shop drawings, especially those who are willing to toil in it. These people can ensure the on-site construction is executed faithfully. Therefore, the Ministry of Education should devote more effort to establishing vocational colleges. During my time at Ho & Partners Architects, there were more than 70 people with clear hierarchy. The boss occupied the inner sanctum while the architects and directors were situated along the window with their assistant and clerks, and then the junior architects and the draftsmen in the periphery. One-fifth of their employees were architects and three-fifths were draftsmen. The staff structure was reasonable and effective at lowering labor costs. As for us, we do not have such a hierarchy. We just have seniority based on when you joined the institute. I am wondering how is the development of vocational education in Taiwan?

PAN: Taiwan has a similar issue. I think traditionally the Chinese people have not held the artisan spirit in high esteem and are not willing to become artisans. Ever since ancient times the Chinese people have believed that constructions are merely the work of artisans. In this respect, the western countries are better. In foreign firms, people who are highly skilled in construction drawings are proud to give technical guidance and certainly do not sell themselves short. They also have valuable employees who specialize in the construction specifications, since many of the construction contracts are based on the construction specification with the assistance of cost estimators. Many great firms attach great importance to cost estimation because designed items can only be realized when the budget is reliable. If such expertise is not valued, then the quality of the overall architecture industry suffers.

"Any high-caliber large design firm has accumulated ample social resources and experience over the years. Nations and societies will make progress only when professional experience can be accumulated and inherited."

"一个有规模、有水平的公司、设计院，其实已经累积了多年的社会资源和经验。专业经验能够积累和传承，这个国家、社会的建筑才会真正进步。"

走向"匠人"精神——专精化的人才发展趋向

崔：我想听听台湾在职业教育方面的情况。记得我们大学毕业刚进入设计院时，设计院有两类技术人员，一类称为建筑师、工程师；一类称为制图员或者技师、高级制图员，来自中专职业教育体系。后来职业教育在大陆被弱化了，可能现在施工单位中还存在一些职业中专毕业的人员，但建筑设计行业中已经没有了，现在大设计院的工作人员以研究生为主。团队中都是新人帮资深设计师画图，每个人的职业目标也是希望成为设计名师，甚至也有以做大师为职业长远目标的。这造成设计院中缺乏把图画得很精到的职业建筑师。我认为大陆的建筑设计水平以及建造成果不十分理想，原因之一是我们缺少一个很稳定的有匠人精神的设计团队，这是一个很大的问题。

我们本土设计研究中心很欢迎擅长做施工图、能够一直认真做施工图的人，有这样的人在现场能确保设计执行到位。所以现在职业教育应该被重视起来，国家教育部应该鼓励多办一些职业学院。早年我在香港何显毅事务所短期工作，他们的事务所有 70 多人，层级清晰，老板坐在最内，建筑师、室主任沿窗而坐，旁边是助理与文员，接着

是次一级的建筑师以及依次而坐的绘图员。真正的建筑师比例大概占有五分之一，有五分之三是辅助绘图员，不仅人员结构区配合理，也有效地降低了人工成本。我们就没有这个层级，大家差不多都一样，只是来的时间长短不同而已。不知道台湾在职业教育方面的发展怎样？

潘：台湾也有类似的问题，我想中国人从古至今对于匠人精神并不推崇，也不太愿意成为匠人，中国人认为建筑都是匠人在做的事。西方国家在这方面做得比我们好。在国外事务所，做施工图的人非常专精，他们以技术指导为荣，也不会觉得矮人一等。不仅有施工图，还有专研施工说明书的人员，因为建筑很多合约都是依据施工说明书，加上估价人员协助完成的。很多优秀的事务所都重视估价，因为估价可靠，设计出来的东西才能符合预算、如实发包。这些人都是专家，如果得不到重视，将直接影响整体建筑产业的质量。

我在国外待了颇久，看到这些问题，所以一直很注重专业人员的培养。设计图、施工图都需要经过品质管理部门的审查、把关；事务所也培养了一个估算部门，现在有三位同事专职，随时与市场保持较好的联系，确保我们做出来的设计可靠，能够发包。设计发展不能到最后再做估价，在过程中的几个阶段，估算组就会提供一些数据、信息，项目依据估算的材料和造价执行，不能任意降低施工标准。

另外，事务所也编制了一个不小的施工监造部门，台北事务所目前约有 200 人，专职监造的同仁就有 60 人左右，多半具有丰富的经验，而拥有 10 年以上工作经验的人员占比极高，他们主要在现场把关。台湾有些公共工程会规定要有现场监造，即使没有此项规定，我们也会向业主要求提供监造，表明监造是品质管理把关的保障，如果业主同意，就会同时支付我们监造费用。这些监造人员经验丰富，能够做到项目的“事先防范，事后检查”，也与设计师保持密切沟通，项目落实起来相对容易一些。很多项目在设计、施工图阶段的时间不充足，常因一些特殊原因需要赶进度，如果有现场监造人员把关，遇到设计图交代不清楚时，他们可以咨询设计人员厘清疑点。业主明白了监造的重要性，多半会给予支持。

崔：目前这些在大陆是缺乏的，同时也是我们这几年一直期待改变的。三四年前，大家研究怎样提升设计质量，近两年住建部领导也比较关注设计，邀请很多专家向政府提交专题报告，其中一项就是有关建筑师的责任制。对于现在大陆推行的五方责任

I worked abroad for a long time so I see these problems. Therefore, I always pay great attention to the training of professionals. At JJP, all design and construction drawings need to be reviewed by the quality control department. Our firm also has a cost-estimation department with three full-time colleagues. They have close contact with the market to ensure our design is feasible for bidding. But cost estimation cannot just be made at the final stage of the design. The estimation team would assist the designers during the design phase so the project can be carried out with an estimate of the materials and cost, which also prevents subsequent lowering of standards.

In addition, JJP has also established a construction supervision department. Currently, we have about 200 people in Taipei and 60 of them are construction supervisors working on site to control the construction quality. Most of them have extensive field experience and a significant portion have more than ten years of experience. Many public projects in Taiwan require on-site supervision, but even if there is no such provision, we still ask our public and private clients to have us provide such service. We make clear that supervision is essential to ensure quality control. Our experienced supervisors take many precautions to prevent errors from happening in addition to inspection work. They also keep close communication with the designers, especially since there is often insufficient time for design and shop drawings. When the design drawings are not clear enough, the on-site supervisors can immediately consult designers to clarify. Clients usually accept the service after understanding the importance of the supervision work.

CUI: This is what we lack in mainland China now, and is also what we have been trying to change in recent years. About 3 to 4 years ago, we started to study how to improve the design quality, and in the past two years, leaders of the Ministry of Housing and Urban-Rural Development have also begun to pay more attention to design. Experts have been invited to submit special reports on related topics, one of which is the responsibility of architects. Under the five-party responsibility system now implemented in mainland China, I think many regulations are unreasonable. For example, since architects have no right to on-site supervision, it doesn't make sense to have the architects stamp all drawings and take full responsibilities for projects. In recent years, architects even have to take on a life-long responsibility after the building is completed. In reality, architects can only be responsible for the design drawings. We cannot take the responsibility if someone changes the drawing without our permission.

Generally, we hope that the design institute can acquire the right to construction supervision and assume overall responsibility. Some firms have started to do management work, and since management is a professional science, they have recruited many young people who majored in project management. Although these young people do not specialize in architectural design, they have very good understanding toward the

制，我觉得有很多不合理的地方。建筑师没有监造权，要求建筑师签字对所有事情负责其实是不合理的，甚至几年前开始施工的项目竣工后，建筑师也要签字终身负责。建筑师只能为设计图纸问题负责，如果有人修改了图纸而我们不知道是没办法负责的。

总体来讲，我们希望设计院能够拥有监造权，称之为设计总包。现在一部分建筑师转做管理牵头人，但是其实管理是一门很专业的科学，后来这些部门也招收了很多工程管理专业的年轻人，这些年轻人虽然对设计不如建筑师了解，但对管理上的基本方法和工具，还是比较了解。另外还有一种情况叫设计牵头的总包（EPC），设计牵头和施工一起统筹管理，同时也负责材料、采购和成本的控制。大陆以前的工业设计院拥有这样的能力和工作模式，一般民用院没有，我们也不熟悉这种操作，真正要做也要准备承担风险。现实点说，我希望建筑师把监造作为一个职业内容，和施工营造商一起控制成本。现在大陆的情况是，比较认真的建筑师如同尽义务一样不惜成本、不断地派人去现场监造，有的业主希望有设计院人员在现场，遇到事情也会征询设计院的建议。但是对于这种监造工作并没一个相应的收费标准，我们只能通过设计上的修改、调整另外收费。所以，交给建筑师监造权也一定要解决服务取费的问题才行。

刚才说到匠人精神，在最近的评审，尤其是建筑传媒奖的评审当中，我看到一批知名的青年建筑师，很有匠人精神，细部完成得很讲究。似乎知名的建筑师本人倒更愿意成为某一类匠人，但是大多数年轻人这个观念还没有转变过来，还在以成为建筑大师为追求目标。

潘：台湾在这方面也未上轨道，有点可惜。许多在大学教学的知名老师，经常上杂志、写文章、发表意见、参加投标评审，但是他们并没有太多的实务经验，常以纯理论的观点，而不是实践的角度去评判一个设计的优劣。我觉得把一个设计项目实践到最后、贯彻始终，是建筑师很重要的一项任务。

建筑评论与媒体责任

崔：此外大陆有一个变化值得说一下，就是建筑评论开始对建筑设计有一些影响力了。很多年来大家都从事设计，不太愿意去做评论，真正的评论一直没有发展起来。最近几年我觉得大陆的学界中有一批很认真、有理论基础的评论家出现了，大概在四十多岁的年轻一代中有几位。我觉得他们在学界的推动作用相当大，可以比较深入地从文化、历史、社会层面，从整个建筑学的发展层

basic tools and methods and of project management. There is another situation called engineering procurement construction (EPC) in which the designer coordinates procurement and construction management. Previously the industrial design institutes in mainland China had such ability while the civil design institutes didn't. We are neither familiar with this operation mode nor capable of taking such high risks. Honestly, I hope that architects can provide supervision service and control costs together with contractors. The case now is that dedicated architects would carry out on-site inspection at their own costs. While some clients would like to have architects on-site and seek advice from the design institute to solve problems, there is no billing standard for the supervision work. We can justify additional charges only through design modification. Therefore, the contractual issue needs to be dealt with as well.

As for the artisan spirit we just mentioned, as a juror in recent awards competition, such as the China Architecture Media Awards, I found a number of young and well-known architects who have the artisan spirit and pay attention to the execution of details. It seems that famous architects are now more willing to work as artisans, but most of the younger generation is still pursuing to become the next 'starchitect'.

PAN: It is a pity that Taiwan is not on the right track as well. Many famous architectural professors frequently publish reviews, offer commentary, and participate in tender assessments. However, most of them do not have too much practical experience and often evaluate design from a purely theoretical point of view, instead of a practical perspective. I think it is important for architects to have carried through a project.

Architectural Reviews and Media Responsibility

CUI: There is a change in the mainland worth mentioning, which is that architectural reviews are beginning to make an impact on architectural design. For years, people have been focusing on doing design work but not very willing to offer commentary, which led to the architectural review field being underdeveloped. However, in recent years several young critics around 40 years old who are dedicated, have solid theoretical foundations and are quite influential have emerged from the academia. They are able to offer criticism from the perspectives of culture, history, society and architectural development. Architectural criticisms from a higher vantage point serve as a very good guidance for those architects who are willing to observe and study. These reviews are working pretty well on evaluating architecture's value, the built process, the design thinking, and the transfer of cultural values among these thoughts.

I do think the readability of design magazines in China is much better than before. It used to be just architects introducing their own projects but now magazines have different styles and include more interviews and commentary.

面去评论一件作品，有了学术的高度，这样的评论对于那些愿意去观察、去读书的建筑师可以产生很好的引导。怎么去看待一个建筑的价值、它产生的过程、建筑师的思考，以及这些思考中文化价值观的转移，所有这些东西都做得不错。我觉得现在大陆杂志的可读性要好过以前。以前都是建筑师自己写文章介绍，现在形式丰富起来，有对谈也有评论。

潘：我无论在学校跟学生讲授，或有机会向媒体发声时，都非常强调建筑评论的重要性。建筑领域这么广，但设计只占整个建筑领域 20%~30% 的比例，还有很多其他发展空间。在国外，大多数一级城市的报纸每个礼拜至少刊登一篇建筑评论，对所在城市或者附近的建筑发表一些观感和想法，同时也接受大家批评与讨论，我觉得这对普及建筑非常有帮助。

第一，可以让社会大众进一步了解建筑。不只是使用者，每个人都会受到建筑的影响，环境、城市也会受到影响。它可以让社会大众对建筑多一些了解和品位，大家的关注会产生一种正能量。

第二，对建筑师会形成一个动力及约束。建筑要存在几十年甚至百年，设计不能没有约束；很多建筑师根本不在乎对环境、对人的影响，只想做出一些彰显自我的设计，所以建筑评论就更显得重要了。台湾到现在还没有建筑专栏作家，很可惜。

崔：这几年在大陆某些建筑成为热点，尤其在网上，但评论多是七嘴八舌的状态，没有专业的专栏作家，有些采访和报导也未必专业，评论所起的作用不大。如果建筑评论可以发挥社会教育、舆论监督的作用，专栏作家独立的身份被确立起来，媒体拥有常设的栏目，这样就可以更好地对社会大众产生影响，偶然的评论不容易引起大家的关注和兴趣。

潘：美国有几位建筑评论家，多半是女士，他 / 她们是非常受社会尊重的。现在我们常常看到有些建筑登报时，报导说首长剪彩、影歌星助阵捧场，照片下会标注出摄影师的姓名，通篇却不会提到建筑师，这种情况非常可笑而不合理。一个电影、一本书、一幅画一定都有作者，建筑却没有作者？建筑对社会的影响是非常深远的，西方在这方面似乎比我们要进步得多。

崔：我们多年前也在抱怨这种情况，有关建筑工程的报导一般就是工程剪彩或者竣工报导，但是多数只报导施工单位却并不提设计单位，这种情况

PAN: I often stress the importance of the architectural criticism whenever I engage the academia or the media. Architecture is a broad field and design only accounts for 20% to 30% of its whole, so plenty of other possibilities exist. In most first-tier cities abroad, newspapers would publish at least one architectural review each week to discuss projects in or around the city. Public comments are very much welcomed. I think this is very helpful for the popularization of architecture.
First, it provides the public more understanding about architecture. The truth is, it's not only specific building users but the general public, the city and the environment is impacted by architecture. Such reviews allow the public to understand architecture better and lead to more public attention, which is always positive. Secondly, architectural review can motivate and restrain architects. Architectural design needs constraints because a building typically last for decades or even centuries. However, many architects do not care about their projects' impact on the environment or people. They just want to design something to show off themselves. Such acts make architectural criticism even more important. It's a real pity that Taiwan hasn't had an architectural columnist yet.

CUI: Several architectural works have become hot topics in the mainland recently. However, most of the reports and interviews, especially those on the Internet, from columnists without professionalism are not necessarily professional and work not very well. If columnists can establish their independent identity and the media introduce regular columns, then architectural criticism can perform the role of social education and public review, thus having a greater impact towards society. After all, it is not easy for the occasional articles to attract much public attention and interest.

PAN: In the case of the United States, several architectural critics, with a majority of them women, are very socially respected. Whereas in Taiwan, when we see an architectural project published on a newspaper, the article would mention the people who cut the ribbon and celebrities attend the celebration. The newspaper photo will also be credited but there would often be no mention of the architect at all. This is really hilarious and ridiculous. Every movie, book, and painting has its author. How can an architecture be built without an architect? The social impact of architectures is very profound. On this issue, western countries seem to be much more advanced than us.

CUI: We also complained about this situation many years ago. News reports of architecture are mostly about ribbon cuttings or project completions, with mostly the construction company named only. Such situation still exists now. Sometimes the name of the design company would be mentioned but not the name of the architect. I am not sure if a disrespect for architects is a tradition among the Chinese people. Meanwhile, I discovered that architects enjoy high status in many other developing countries.

现在依然存在，即便有时会提到设计单位，也很少提建筑师。不知道这是不是华人圈子的一种陋习，不太尊重建筑师的劳动。我发现在其他许多不发达的发展中国家，建筑师的地位都很高呢！

建筑设计事务所的社会影响

崔：对于什么规模才是一个事务所最理想的大小，曾经在跟国外同行交往当中被问起事务所的大小时，我说我们院（中国建筑设计研究院）拥有约8000人，他们表现得很惊讶。8000人其实涵盖了多个行业，包含建筑、城市规划、环境景观、历史研究、市政、城市公共卫生和咨询等众多的人员，真正的建筑工程设计人员只有1800多人，其中建筑师约五六百人，最后说到跟我工作的团队大约有50人时，人家才点头认同，觉得这个是事务所合适的规模大小。这个现象很有意思。其实从一个人的掌控力来讲，我觉得30多人就达到极限了，所以当团队到达50人时，我将它分成了几个小团队，每个小团队由一人负责，小团队内有十几个人，我觉得这样开展工作也是比较有效的。
我们的工作基本上是采取师父带徒弟的模式，团队中的项目各阶段我都会参与其中并提供一些意见，有些项目自己也会直接负责并管理。但是总体来讲，进入建筑设计的阶段时，团队自身要有执行力，当然不少项目状态也不稳定，所以经常是一组人穿插多个项目。

潘：一个有规模、有水平的公司、设计院，其实已经累积了多年的社会资源和经验。如果因为主事者年纪大了退出公司，这些经验和资源便失传就太可惜了。台湾过去也经历过非常蓬勃的发展期，很多

The Social Impact of Architectural Design Firms

CUI: What is the ideal size for a design firm? I was once asked by foreign peers about the size of our office. I said that our institute (China Architecture Design & Research Group) has about 8,000 people. They got very surprised. The 8,000 number actually covers a wide range of disciplines, including architecture, city planning, landscape architecture, historical research, municipal administration, public health and consulting. The architectural department is comprised of no more than 1,800 designers, of which 500 to 600 are licensed architects. It was not until I told them I had about 40 to 50 people in my team that they began to nod in agreement that such a number was an appropriate size for an architectural firm, which is very interesting. Of course, in terms of one's management ability, I think more than 30 people would have reached a limit. So when my team reached up to 50 people, I divided it into several small teams, each with about 10 members and a leader. I think it is more efficient to work in this way.
The teams all adopt an apprenticeship mode. I will participate in every phase of all the projects and provide advice. I would also directly manage some projects. But generally speaking, when a project enters the architectural design phase, the team itself needs to be capable of execution. Of course, not all projects progress steadily so each team usually has several projects simultaneously.

PAN: Any high-caliber large design firm has accumulated ample social resources and experience over the years. It is a shame to lose these resources just because the founders or leaders have retired. Taiwan has enjoyed a very prosperous period in which many established architects did very well, but their experience did not pass on at all when their firms simply dissolved after their retirement. Nations and societies will make progress only when professional experience can be accumulated and inherited.
I am quite fortunate in receiving a rather complete architectural education so I wanted to bring my experience back to Taiwan. When returning to Taiwan, I started from zero but when the firm started to grow, I consulted an American construction management consultant about the ideal firm development. He said that one person can manage about 25 people at most, so I set that as the target size. Back then Taiwan was developing its biotechnology and IT industries, and our growing firm was fortunate to participate in these projects. Our staff grew accordingly from 10 to 20 and then 50 people. Later on we established an interior design company and our structural engineering department became an independent company of about 20 to 30 people. When our long-term clients asked JJP to head across the strait with them, we set up offices in Shanghai and Xiamen. Having the office grow into an organization of nearly 300 people after 35 years is definitely beyond the original plan.

CUI: Mr. Pan, how many people do you know in your firm?

知名的老建筑师做得非常好，但是完全没有传递下去，人走了事务所就解散了。专业经验能够积累和传承，这个国家、社会的建筑才会真正进步。
我个人是很幸运的，接受的建筑教育过程也相当完整，希望能够把个人的经验带回台湾来。回到台湾时一切从零开始，慢慢有了一点成长后，就思考应该怎样去发展，当时曾咨询过一个美国建筑管理顾问。他说，按照面面俱到的理想状态，一个人能够管理的极限大概是 25 个人，我们便以 25 人为目标。发展到二十几人时，台湾刚好发展生物科技、IT 产业，事务所有幸参与这些项目，人员从二十几人一路发展到 50 人，后来加上工程配套、成立室内设计公司，接着结构部门也发展成一个 20~30 人的结构设计顾问公司。长期合作的业主也在业务西进时，希望我们能够跨海协助他们，所以又相继成立了上海、厦门分公司。成立 35 年的事务所发展成

PAN: I recognize almost all the faces. When the firm had 100 to 200 people, I could remember everyone's name. Now there are some people I only see once in a while, especially those who are sent directly to the construction site after joining, so it's hard to remember their names. But with the exception of those 20 to 30 people, I can remember the rest of the staff's names. Since we work hard together for long periods of time, I would often ask about their well-being and even their families. Our office has a real familial atmosphere. We organize various activities, including on-site visits and an annual domestic trip. We also welcome our colleagues to bring their families to our year end party, which gives me the chance to meet them. I have witnessed some of my colleagues' children grow from babies to graduate or doctoral students. It is very important to have the firm feel like a family. Of course, there is a negative side which is that sometimes it is difficult to deal with problems objectively.

I began to think about passing the torch ever since I was 60 years old, just when the partnership type of firm ownership became legal in Taiwan. Now the firm has seven partners which will increase as the firm continues to grow.

CUI: Nowadays, the internet is a popular topic. People in our institute have been mentioning to me about how architectural design can adapt and adopt Internet Plus. Also, some people at online forums have talked about how the construction industry will be like a home delivery service in the age of Internet Plus. They imagine that in the future clients can order from home, choose different institutions to develop different schemes or different parts of the project, and then put them together to form an entire building. On this matter, my opinion is not that optimistic. Unlike cooking or rendering, given the complexity of architecture, which involves huge investments and responsibilities, I don't think a building can be finished without face to face interactions.

As an ancient industry, it is hard to predict the various forms architecture might take in the future. I think the architecture industry can continue to develop under the current system, at least in the foreseeable future. Cui Kai Studio has been established since ten years ago, and was renamed the Land-based Rationalism D•R•C in 2014. In the future, the office can accommodate other studios to form a design brand and continue to develop.

PAN: Passing on the torch is really an important issue. When the firm converted from sole proprietorship to a partnership, I suggested to change the name of the firm. But the other partners thought that JJP had finally accumulated some recognition after 20 years of development, and if the name was changed, people would then need time to reacquaint with us. Therefore, we designed a logo of our firm with just the letters JJP. Our hope is that one day even if people don't recognize the full name, they will know what JJP represents. In other words, the logo will become a symbol of professionalism and a culture that continuously strives to make a positive social impact.

一个近 300 人的组织，是完全不在计划内的。

崔：潘先生，你大概认识你公司多少人？

潘：300 人面孔差不多都认识，早期 100~200 人的时候公司每个人的名字我还可以叫得出来。现在有一些人，尤其是面试后被直接派往现场工地的员工，偶尔才见到他们一次，很难记住。所以现在除了 20~30 人可能记不住名字，其他的都可以。因为大家一起卖力做事，我有机会跟不同组员讨论时，多半都会进一步了解一下他的状况，甚至他的家庭。我们公司的氛围就是非常家庭的感觉。我们会举办各种活动，包括工地参观及每年一次的员工旅游，年会我们也会欢迎公司同仁带家眷一起参加，所以有机会认识他们的家人，甚至看着同仁的小孩从抱在手上，到现在有些已经读研读博了。家庭般的感觉是维持公司工作水平的一个很重要的因素，当然负面的影响可能就是太讲人情，导致一些事情不能铁面无私地去处理。

我 60 岁时就在考虑传承，那时联合事务所制度在台湾刚被允许，现在已经有 7 位合伙人，以后还会陆续发展，加入新合伙人。

崔：现在互联网是时下的新话题，我们院里也有人跟我谈论，如果以后建筑设计变成“互联网 +”的形式，设计院还能生存多久。也已经有人在论坛上谈论建筑行业的“互联网 +”，如同送货上门，如果将来业主可以在家点餐，就可以选择不同的机构个体做不同的菜或者一道菜不同的部分，最后拼合起来，形成一个建筑。对这件事情我的看法并没有那么乐观，因为建筑行业技术复杂，牵扯到巨大的资金投入和责任，我觉得还不能像做饭一样，也不像画效果图无需见面就可以完成。

作为一个古老的行业，建筑未来是不是会有多种呈现形式，现在还很难说清，我觉得在我可见的范围内建筑行业是可以持续发展下去的。崔愷工作室开办了十年，后来名字改为本土设计中心，以后多个工作室也可以在这个中心发展起来，形成一个设计品牌，也是可持续发展下去的。

潘：传承真的是个重要的课题，联合事务所刚成立的时候我提议更换名字，但是其他几个合伙人认为事务所积累 20 年的知名度，如果换了名，大家得重新熟悉……所以，才找了专人设计事务所的 logo——JJP。可能以后大家不知道它的全名是什么，但是看到 JJP 就知道它代表的是哪个事务所，希望可以成为一种专业符号、一个文化象征，持续发挥正向影响力。

Taoyuan International Airport MRT Stations

台湾桃园国际机场捷运车站

2006–
Taipei
8 MRT Stations

Stations as Part of Urban Culture
车站，城市文化的一部分

For travelers, rapid transit stations linking airports and cities are nodes where the pleasure of starting journeys is celebrated. Considering the significant influence of stations on the public environment, our design started from the question "what impression shall be created for travelers when entering or leaving the station platform?" and demonstrates itself in eight stations on two different sections of the transit line.

对旅客而言，认识一座城市从车站开始，车站兼具便捷交通及传递城市形象的任务。车站同时也是在地居民的生活中心，对公共环境的影响深远，因此站体建筑应积极串联、整合带动周边城市的成长。作为基础建设的车站，对市民与访客来说是城市生活的一部分，不只是实体建设，也塑造了具有文化价值的公民意识与认同感。面对深入生活的建筑类型，在设计策略上，以"为乘客在进出车站时留下印象"为思考方向，逐一展现在机场捷运系统两路段的八座车站上。

Taipei Section—Station A1 to A6
台北区段—A1 至 A6 车站

This section consists of one underground main station and five elevated stations. The design emphasizes its coherent structural system. As the gateway to Taipei city, station A1's wide interior spaces interpret the departure and arrival of a grand tour. Other stations are actively linked and echo the individual characteristics of the surrounding neighborhoods.

A1 station design architect: Maki and Associates;
Collaborating architect: JJP
A1 站设计建筑师：日本桢综合计划事务所；
合作建筑师：JJP

Located next to a riverside park, station A2 utilizes a weaving tubular structural frame that renews the relationship between the city and the river. Its curved roof with a thin metal skin is lifted at a slight angle off the horizon to create a gesture simulating a plane taking off. Situated in an industrial park, station A3 uses a steel structure that aesthetically responds to local industrial development. Stations A4 and A5 are located in dense urban areas, and station A6 is at the periphery of an urban area, with their individual designs determined by the neighboring contexts. Incorporating a common theme of fenestration design that guides the direction of travelers, the stations instill a sense of movement and integration of various perspectives and activities from the surrounding environments.

此路段为一座地下总站与五座高架车站，强调以脉络一致的构造系统来建构车站本体，起始的 A1 车站作为台北市的门户，以宽阔空间诠释壮游旅程的出发与抵达，跨都会区沿线各站则呼应城市的不同区位特性。

位于河滨都会公园旁的 A2 车站，以筒状编织结构强调水岸视野，除了联结城市与河道的亲近关系，也借由流畅的弧线造型，隐喻起飞出航之意象；坐落于产业园区的 A3 车站则以钢结构美学表达地区的机械工业轨迹；位于街道窄小局促市区内的 A4、A5 车站，以及都会与丘陵交界的 A6 车站，在配置设计上皆顺应环境涵构布置站体，进而以外墙开口立面造型呈现进出人流动线，创造出带状流动意象串联周边城镇、绿地广场及山景。

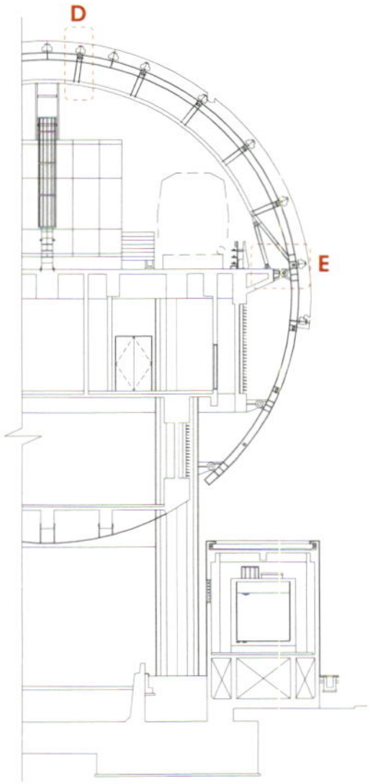

Section steel frame
钢架构剖面

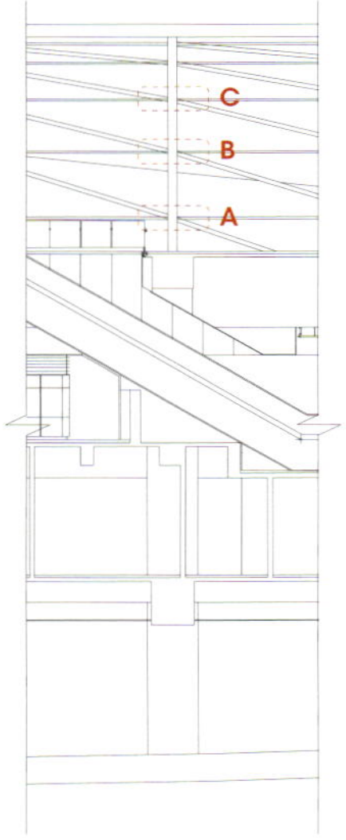

Partial section
部分剖面

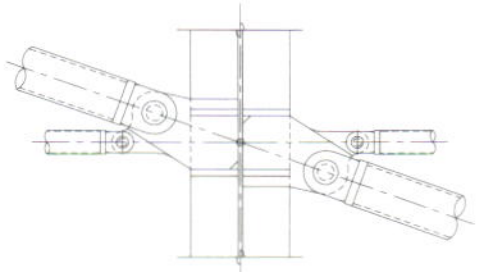

Detail A
细部 A

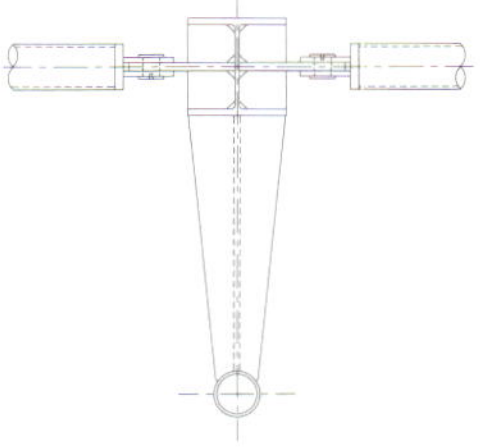

Detail B
细部 B

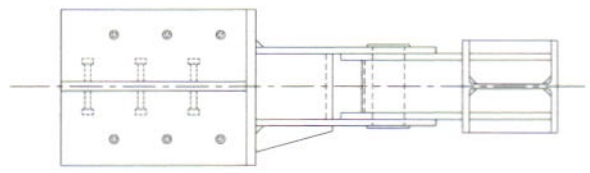

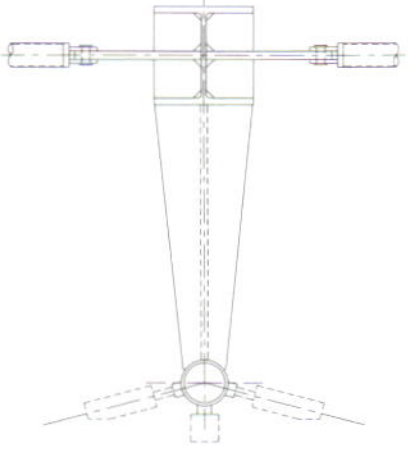

Detail C
细部 C

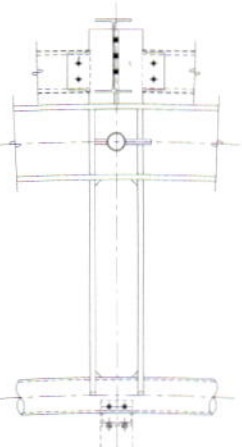

Detail D
细部 D

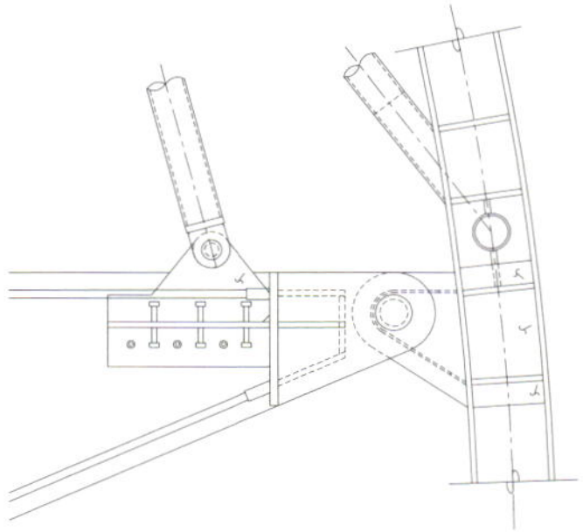

Detail E
细部 E

A2 Detail 细部

A2

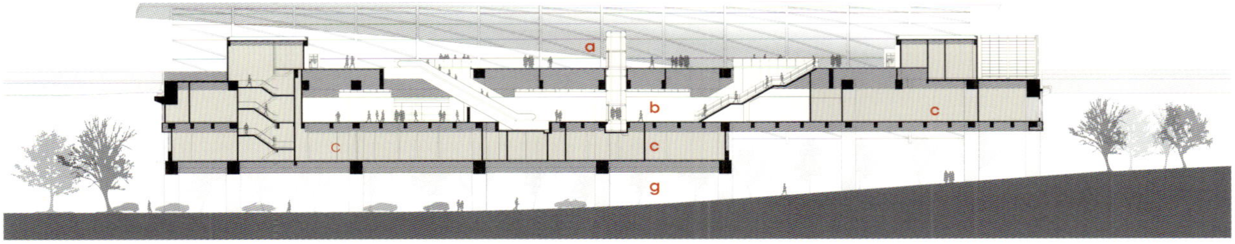

a Platform 月台
b Concourse 大厅
c Mechanical room 机械室
d Connection bridge 连通桥
e Elevator lobby 梯厅
f MRT O5 station lobby 捷运橘线 5 号站大厅
g Dike service road 防汛道路
h Dike 堤顶

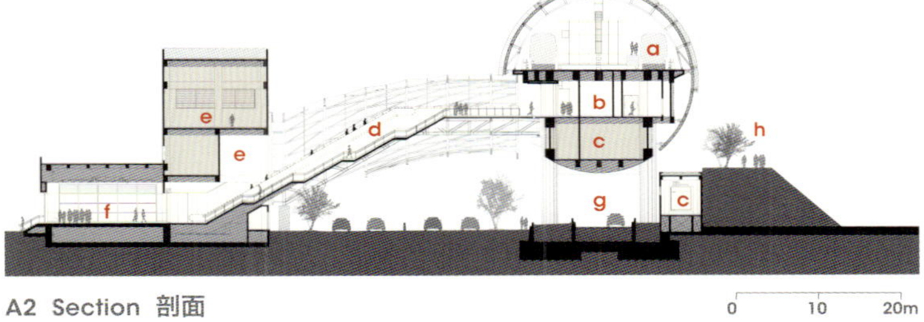

A2 Section 剖面

A2

A4

A5

A3

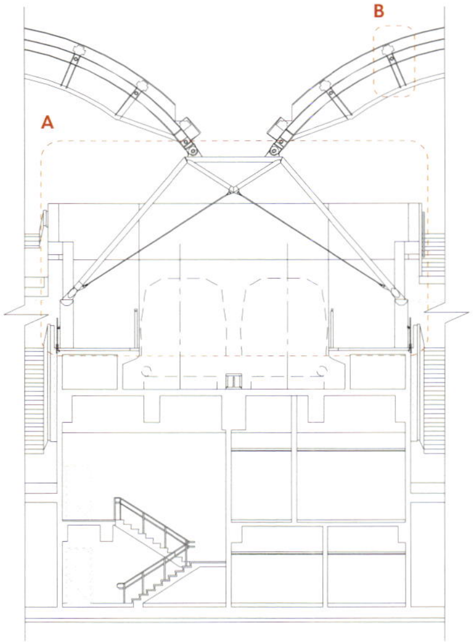

Partial section at central tracks
中央轨道部分剖面

A3 Detail 细部

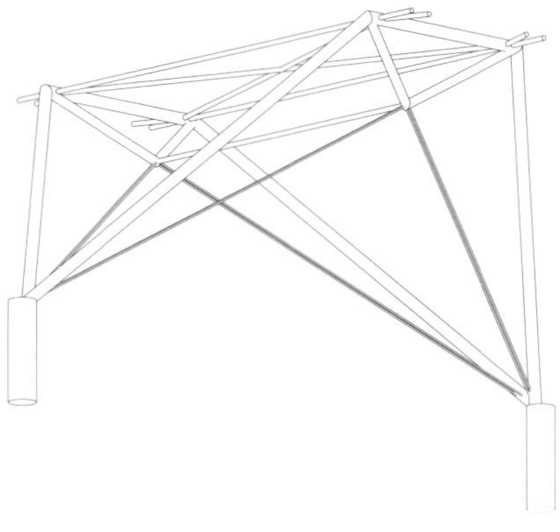

A. Axonometric transitional frame
A. 过道结构等角

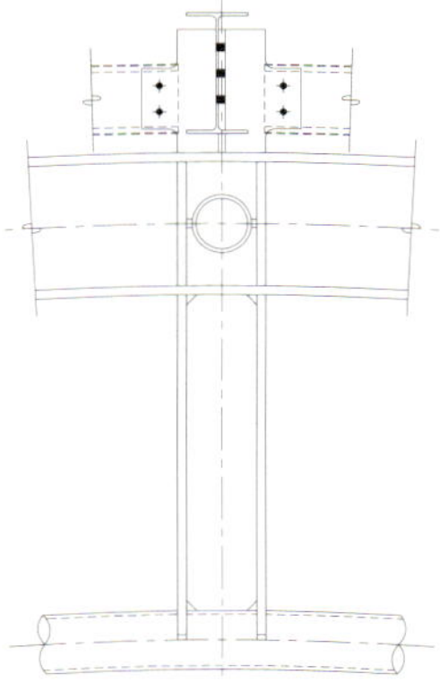

B. Detail
B. 细部

A23

Taoyuan Section—Station A22 to A23
桃园区段—A22 至 A23 车站

Located on the opposite side of the urban area, stations A22 and A23 attempt to reflect on the local geographical characteristics from an earlier era when ponds and hydrophyte vegetation were still lavish. In order to minimize traffic and environmental impact, only the entrances and air shafts of station A22 are located above ground. The two entrances of station A22 and A23 use hydrophyte vegetation as a design image. Since station A23 is situated on the south side of Chungli Train Station, the two stations will be connected to establish a convenient transit environment for passengers. Serving as a focal point in a dynamic city, the station will act as a connector that welcomes vibrant energy, active lifestyles, and culture from all directions.

位于都会区另一端的桃园段 (A22 至 A23 车站)，设计主题来自桃园地区特殊的地景元素——埤塘。A22 站位于邻近中坜市老街溪侧，为了降低对都市环境、景观的冲击，仅留车站出入口及通风井等设施于地面之上。车站两处出入口屋顶特别以埤塘水生植物——台湾萍蓬草作为设计意象；A23 站与台铁中坜站共构，两铁无缝衔接换乘。以流动性的汇集作为设计发想，将文化之流动透过“交、汇”的概念，串接人文色彩及多样化的活动汇流于一交点。完工后不仅可改善周边城市景观与环境，也能以流畅空间与地标性，成为都会生活与文化的新焦点。

A22

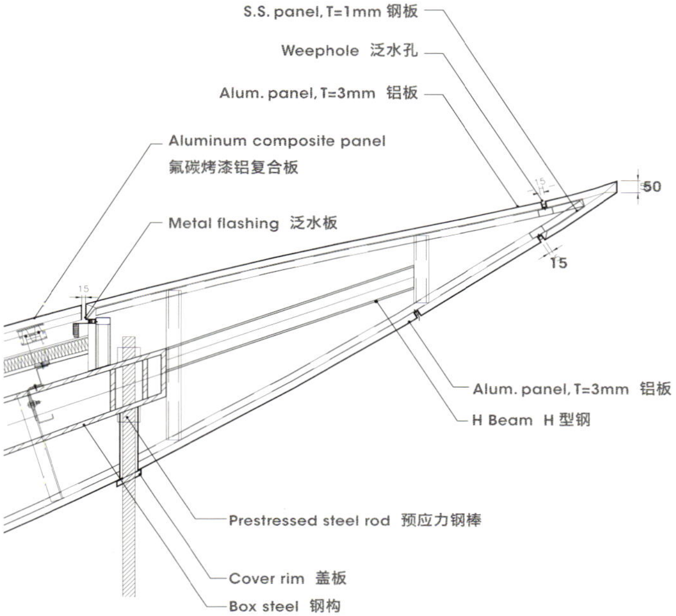

A22 Station curved roof panel detail
A22 站体曲面屋顶细部

A22

When The Bells Toll: Chung Yuan Christian University Series
当钟声响起：中原大学校园建筑系列

1

Chang Ching Yu Memorial Library
张静愚纪念图书馆 *

1983–1985 | 12,000 m²

2

Gymnasium & Indoor Swimming Pool
体育馆及室内游泳池 *

1987–1991 | 10,580 m²

Joint Venture Architect: Chiu Hwa Wang
合作建筑师：王秋华

* 注: Joint Venture Project 合作项目

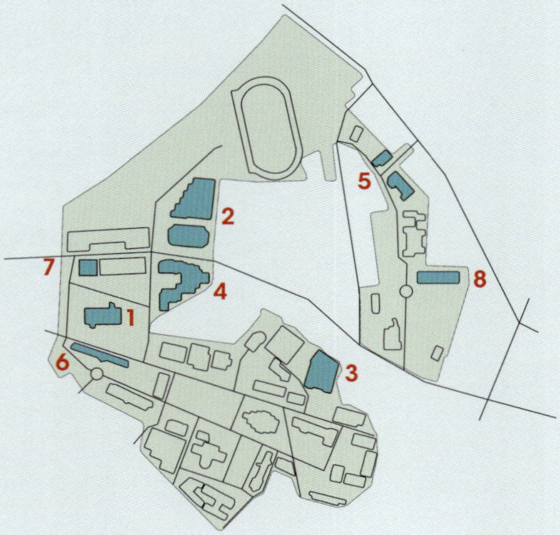

1 Chang Ching Yu Memorial Library
张静愚纪念图书馆
2 Gymnasium & Indoor Swimming Pool
体育馆及室内游泳池
3 Engineering Building 工学馆
4 Holistic Education Village 全人教育村
5 Student Dormitory: Re Cheng Hall
新中北学生宿舍：热诚楼
6 Huai En Hall 怀恩楼
7 Bio-science and Technology Building
生物科技馆
8 Industry-academia Cooperation Building
产学大楼

■ Designed by JJP JJP 设计校舍
■ Campus area 校地范围

4

Holistic Education Village
全人教育村 *

1997–2004　　23,430 m²

3

Engineering Building
工学馆 *

1994–1996　　18,580 m²

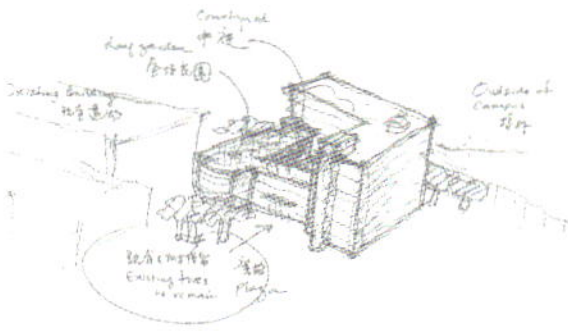

5

Student Dormitory:
Re Cheng Hall
新中北学生宿舍：热诚楼

2013–2015　　27,727 m²

JJP has designed several buildings over the years for Chung Yuan Christian University (CYCU), including the library, gymnasium, indoor swimming pool, engineering building, Holistic Education Village, student dormitory, and an Industrial-Academic Cooperation building. All of the architectural additions were gradually completed alongside the expansion of the campus. Under a complex site situation, the axis of the campus circulation is reinforced by the construction of new buildings. In addition, the continuous public areas provide teachers and students the space for leisure and communication that define campus life.

JJP 团队先后为中原大学设计了图书馆、体育馆、游泳池、工学馆、全人教育村、新中北学生宿舍及产学大楼等单体建筑，所有建筑配置伴随着校区扩增的过程渐进完成。在校园地界复杂曲折的情形下，借由陆续兴建的各栋校舍，协助校园建立自明性的轴线，创造连续的公共场域，让中原师生拥有停歇、活动与交流的校园生活场所。

Chang Ching Yu Memorial Library

张静愚纪念图书馆

1983–1985
Taoyuan
12,000 m²

Pioneer of Taiwan Passive Green Architecture
翻开台湾被动式绿色建筑的首页

The project, which is a 400,000-volume capacity library with a 750-seat, open-shelf reading room, is not only a pioneer of passive green design but also helps position the main axis of the west side of the campus. A multi-directional plaza entrance is constituted by the orientation between the library and an existing building, a two-story classroom building which was CYCU's first building and was preserved and renovated by JJP. This axis sets the basis for the next 30 years of campus development. Since the entrance level of the existing building is two meters higher than the site, a two-level entry plaza was created to suit the natural grades. The upper plaza is approached from the existing Huai En Hall at the south and leads to the library itself. Each floor is linked by open stairs and void spaces for visual interest and to facilitate natural ventilation. The lower plaza is approached from the dormitory and eastern campus area to the east, and leads to non-access-controlled ancillary facilities which include a 150-seat conference hall, five temporary classrooms, the technical services workroom, a 750-seat study hall in the basement and the student lounge. Besides sunken gardens, the basement has an outdoor terraced classroom on the north which hosts activities, performance, teaching, and also offers a view while facilitating the spread of light and ventilation. The outer walls of the main building utilize an insulating air layer with stepping overhang to help introduce fresh air from the lower part of the building, and then discharge hot air at the top, which allows it to become a passive green building with natural ventilation, providing a cozy reading environment for the users.

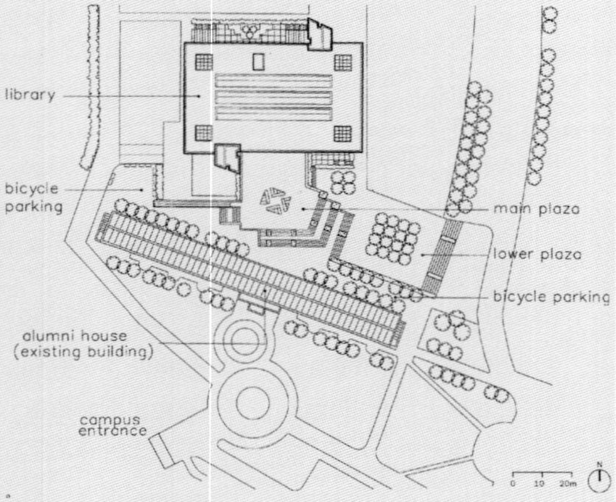

项目为台湾被动式绿色建筑设计的先锋，并为中原校园定位出西侧校区的主轴，其布置走向与创校时首栋二层建筑物怀恩楼，围塑出一个多面向枢纽性的入口广场，树立了往后 30 年校园发展的基础。由于保留并更新的怀恩楼地面层较图书馆基地高 2 米，设计上利用地形高差创造双层入口：上层广场南端由怀恩楼导入二至五层的总图书馆，可藏书 40 万册，为当时台湾首见之开架式书库，含 750 席开架阅览空间，各层分区挑空以提供流畅的视野与自然通风；下层广场则于东端通往教学区及宿舍区，导入开放区域。此区含 150 席会议厅、临时教室、数据处理工作区、750 席地下自修室与学生休息厅；地下室前后均设有地下花园，北面则有户外阶梯教室，除提供活动、表演与教学外，亦提升通风采光能力及景观效果；主建筑外墙皆增设保温空气层，配合在建筑逐层外挑的下方，引进新鲜空气，再由室内挑空空间顶端抽风排出热空气，让师生们可以享受舒适的阅读环境。

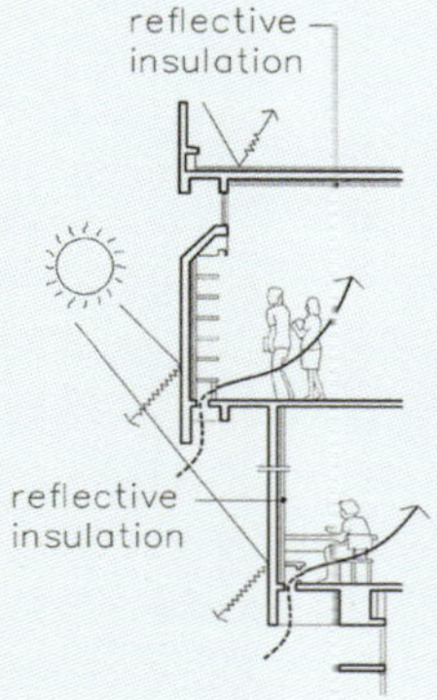

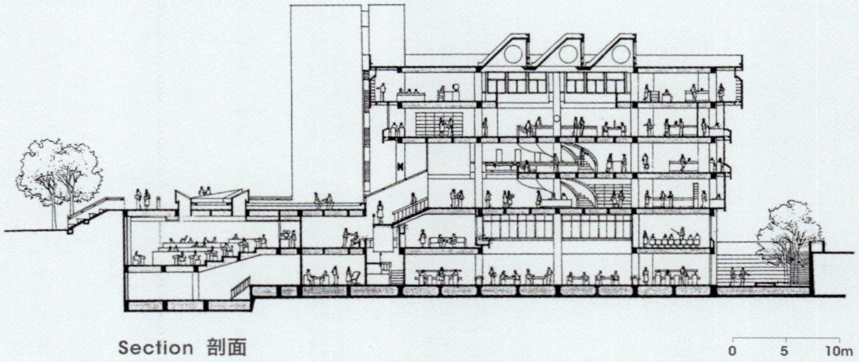

Section 剖面

Holistic Education Village
全人教育村

1997–2004
Taoyuan
23,430 m²

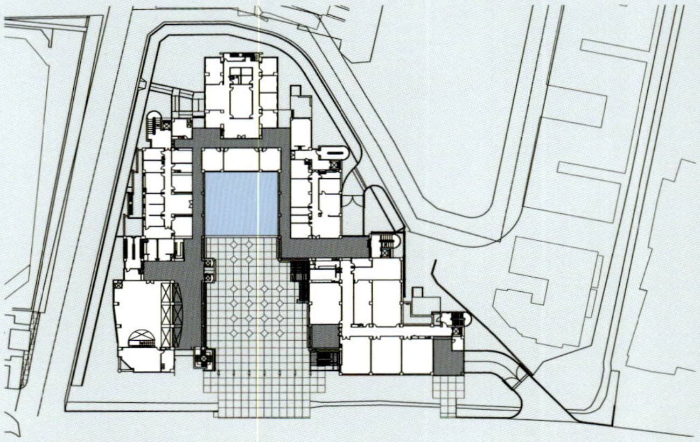

Site plan 总平面

A Harmonious Village
谐和之村

Since the 1990s, CYCU has prided itself in providing "Holistic Education" that addresses the needs of "body, mind and soul." The irregular site borders local streets on two sides and the university's main thoroughfare on the third side. To respond to the complex environment and the west side axis of the campus, the project is configured through a building group of different heights. The north wing is 5 stories tall and contains a performance hall, an art gallery, a conference center, the university archive and the alumni room, plus a health center and guidance suite which looks after the "body and mind"; the 3-story central wing accommodates a concert hall/chapel, the University Chaplains' quarters and fellowship halls to cultivate the "soul"; the 9-story south wing houses the College of Humanities and Education, symbolizing the source of wisdom and compassion. The roof of the chapel is designed in the form of an inverted cone encased in a circular glass clerestory, allowing 360 degrees of light from above, and signifying the Bible's main theme that "The Word dwells amongst us, full of truth and blessing," which becomes the best interpretation of "holistic education."

中原大学自 1990 年开始推行“全人教育”的理念，希望借由“身、心、灵”的培育及“天、人、物、我”的平衡，塑造理想的学习环境。不规则的三角形基地，两面临校外土地，一面临校内南北主要通道。为呼应复杂的基地肌理和西侧校区的南北主轴，项目延伸此轴线配置，以三栋高低不同的建筑群组成鸡犬相闻的教学村落，深化西侧校区的空间层次；北栋 5 层，含演艺厅、画廊、校史室、校友室、国际会议厅等公共设施，与兼顾“身、心”的保健与心理辅导中心；中栋 3 层，含音乐厅兼礼拜堂、校牧室与团契活动室，为培育“灵”的场所；南栋 9 层，则供人文与特殊教育学院使用，代表智慧与人文关怀的泉源；音乐厅 / 礼拜堂天圆地方的造型，圆顶 360° 引入和煦的天光，表达《圣经》中“光来到世间，充充满满有恩典与真理”，光照世人也祝福世世代代的学子们，成为全人教育的最佳诠释。

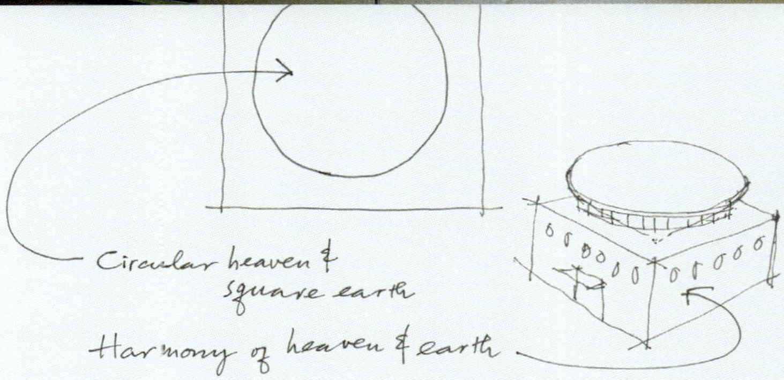
Circular heaven &
square earth
Harmony of heaven & earth

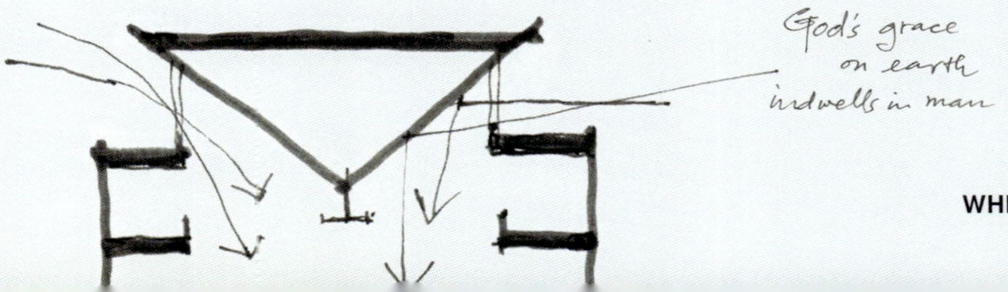
God's grace
on earth
indwells in man

Student Dormitory: Re Cheng Hall

新中北学生宿舍：热诚楼

2013–2015
Taoyuan
27,727 m²

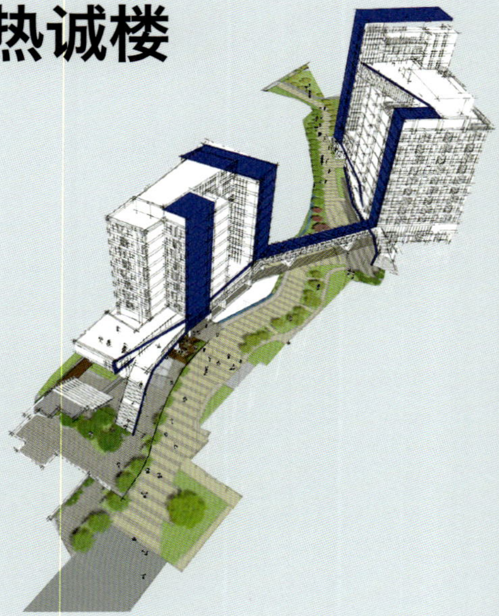

A Blue Ribbon of College Life

蓝色的校园生活回路

Situated at the east side of the ring-shaped campus, the new dormitories define an open area with the existing dormitory through the use of a connecting corridor. The buildings also connect the north side of the stadium with the south side of the campus ecological pond. The use of landscape elements such as an eco-pool, creeks, special planting area and paths, all help to unify the scattered spaces created by the expansion of the school. The school colors are employed throughout building and the public space on the lower floors. The corridor not only extends the feel of the outdoor experience into the interior public space but also protects students from the rain and wind of the northeastern monsoon. By using distinct color blocks, different functions of the indoor spaces acquire clear visual identities. Study rooms that require minimal disturbance are thus placed at the top of each building to provide views and a quiet reading environment. At night, the lighthouse-like visual effect also signifies the rise of the eastern campus of CYCU as an emerging landmark.

新建宿舍位于环状校园东侧的新中北校区，通过连廊设计与原有宿舍围塑出开放广场，并运用校区内的生态水池、水渠、特色树区及景观步道等天然资源，连接北侧体育场、南侧埤塘及周围零散的建筑群落，形成延续的校园生活回路；连廊设计搭配中原大学标志色，贯穿低楼层的公共空间和原有宿舍，延伸户外场域至内部空间，同时减少中坜冷冽的东北季风影响；利用鲜明色块区划不同室内空间类型，兼具穿透性及功能区隔。另将需安静的自修阅读空间集中至顶层，提供良好视野及阅读环境，于夜晚形成如灯塔般的效果。新宿舍除提升了学子们的生活质量，也扩充全人教育的完整版图，为中原大学东侧校园增添了辨识度与亲和力极佳的地标。

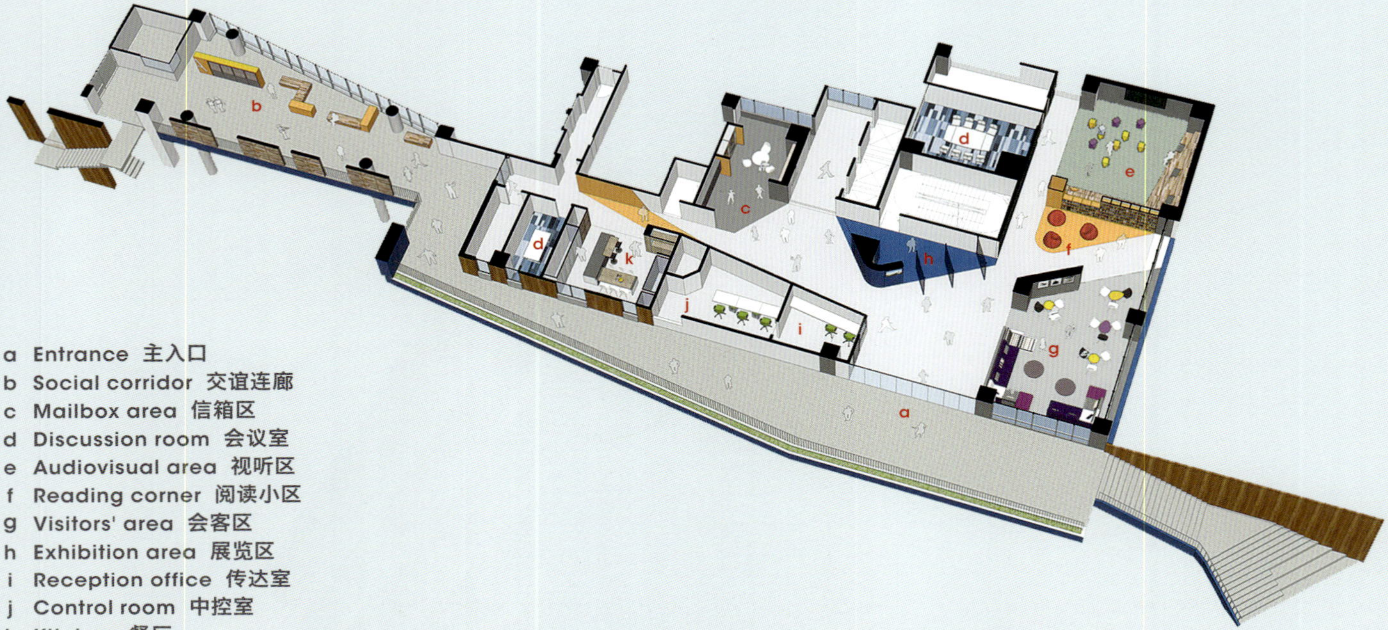

a Entrance 主入口
b Social corridor 交谊连廊
c Mailbox area 信箱区
d Discussion room 会议室
e Audiovisual area 视听区
f Reading corner 阅读小区
g Visitors' area 会客区
h Exhibition area 展览区
i Reception office 传达室
j Control room 中控室
k Kitchen 餐厅

2F Interior diagram 二层室内示意

b

e

f

g

h

k

1

2

3

1 Relaxing area 休闲区
2 Lounge 自修室
3 Light tower 光塔

Mackay Medical College
马偕医学院校舍

2010–2012
New Taipei City
53,366 m²

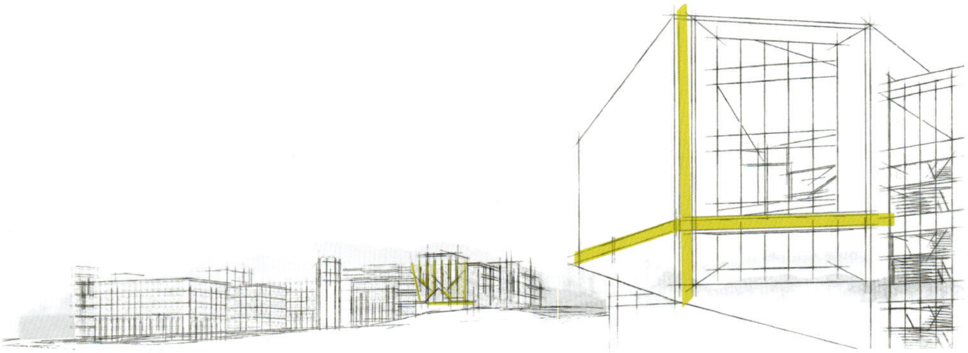

Multiple Layers of Open Space
多层次的开放空间

"Rather burn out, than rust out" is the motto of Dr. Mackay. In honor of his legendary self-sacrifice for the sake of gospel in Taiwan, the design alludes to his spiritual philosophy. The scattered campus planning is in response to the Bible verse "I am the vine, you are the branches," while the Student Activity Center signifies the cornerstone of the church, and a burning thorn-tree is transformed into an iconic green wall for the exterior design of the Academic Building. Situated in a coastal area, the complex adopts a cluster layout to block out the unpleasant northeastern seasonal wind in the winter. Corridors connect the existing and new facilities to create multiple layers of open spaces and to provide a comfortable learning environment.

"宁愿烧烬、不愿锈坏"是来自马偕博士的精神内涵。为纪念他一生在台湾宣教、医疗、教育发展领域的无私贡献，并传扬基督救世之福音，在设计操作上，将马偕博士誓言纳入建筑意象，以植生墙象征"焚而不毁"、以展演厅象征"教会盘石"、以宿舍配置象征"葡萄树与枝子"。此外，由于地处滨海地区，通过"体量围合"及"风雨走廊"的配置策略连接一、二期各栋建筑体，降低冬季东北季风影响，最终目标在于创造出多层次的开放空间及舒适的学习环境。

Social Corridors
走廊上的社群联结

Since the most poetic and unforgettable memories often take place in the school corridors, the three new buildings are connected through corridors that provide a comfortable environment and opportunity for interaction.

Academic Building: An iconic green wall forms the front façade of the building. Corridors are used in-between the existing and the new buildings with several communal spaces inserted along the route to enhance accessibility and user interaction.

Student Dormitory: Responding to the site condition, the dormitory is broken down into several volumes that gently follow the sloping topography, with corridors connecting each area. Each volume is distinct yet connected. Spatial hierarchy is created through various levels of privacy and all housing units are oriented to face north and south to optimize natural light and ventilation.

Student Activity Center: To increase accessibility, several entrances of the building are integrated with the natural topography. The diversity of spatial types not only enriches campus life, but also ensures that the activities of the school are not affected by seasonal weather.

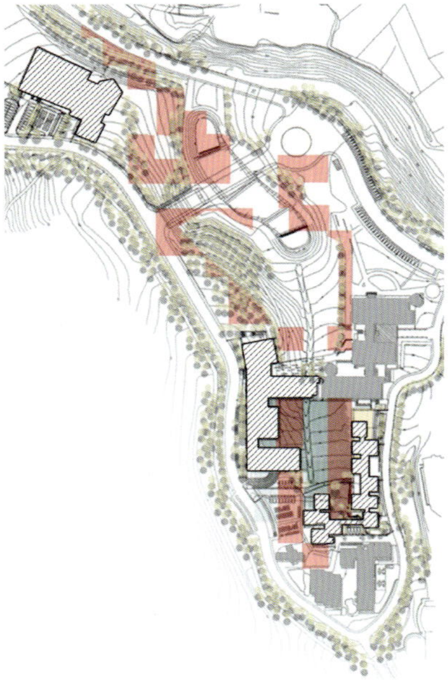

■ Not suitable for building (over 30% slope)
限建地块（坡度30% 以上）

— Connecting corridors 廊道
— Winter wind 冬季风向

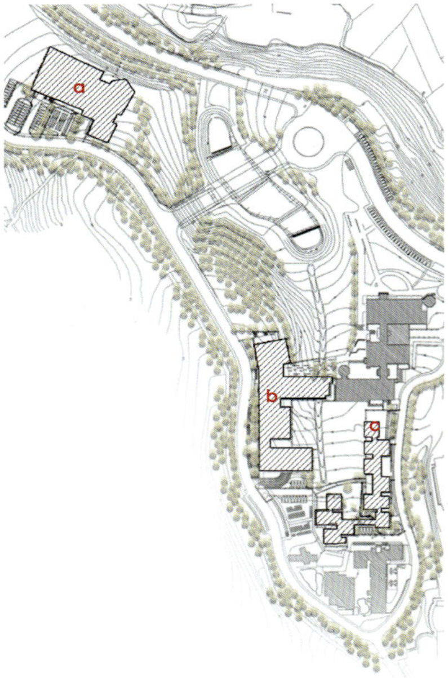

■ Existing buillding 现有一期校舍
▨ New buillding 本期校舍

a Community center 多功能活动中心
b Academic buildings 教学研究大楼
c Dormitory 宿舍

校园最具诗意且令人难忘的回忆，常是在走廊上发生的故事，因此新建三类建筑群之间以通廊串联为整体，兼顾舒适的物理环境，提供交流的机会。

教学研究大楼，其地标性植生墙与一期大楼新旧空间整合，各栋间以回廊串联并设计交谊停驻空间，以增进师生互动；学生宿舍，以家庭单元为群组，有层次地塑造由小至大、由私密至开放之空间架构，住宿单元以南北向配置，兼顾节能、通风及采光效果；多功能活动中心，运用地形高低差整合各空间出入口，增加可达性。多样化的空间类型不仅丰富了校园生活，也确保校内各项活动不受季节风雨影响。

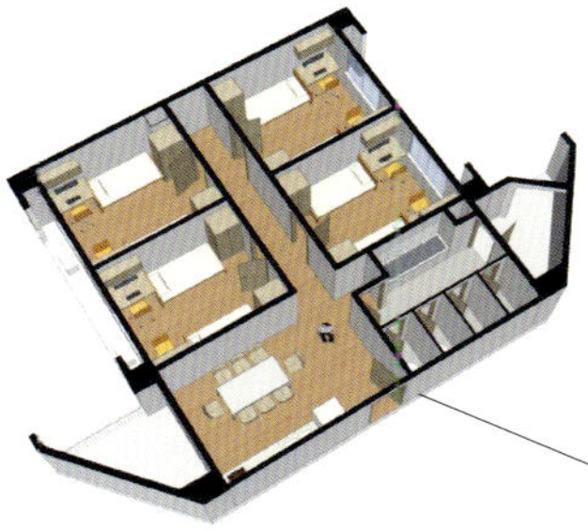

Typical unit 住宿单元

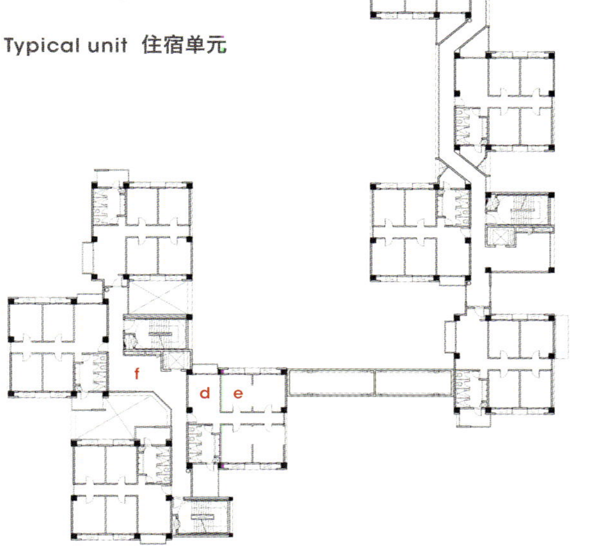

Dormitory plan 宿舍平面

d Living room 客厅
e Bedroom 卧室
f Meeting room 交谊厅

1 Auditorium/Chapel 展演厅 / 教堂舞台
2 Gymnasium 多功能活动球场

1

2

Xinglong Public Housing Block 1

兴隆公共住宅一区

2012–2015
Taipei
25,142 m²

A Public Housing with Universal Design
彰显通用设计精神的公共住宅

Commissioned by the Taipei City Government, the Xinglong Public Housing Block 1 is a rental-only development, and constitutes part of the Ankang Regeneration Project. Located near Muzha park, the project comprises of 272 residential rental units with 19 stories above grade and a common basement. A range of universal design strategies allow for generous accommodations and equal access for all ages and abilities in order to realize social justice.

项目缘自台北市政府 2011 年起推动“只租不售”的公营住宅，属于“安康公营住宅旗舰计划”一区，紧临木栅公园，为两栋 19 层，且地下层相连的住宅大楼，共计 272 户。整体规划配合基地限制打造与环境融合的混居小区，并借由通用设计的理念彰显居住正义精神。

Extend the Civic Life
延续市民公共生活

Strategic planning was essential to transform the site's 6-meter elevation difference into an advantage. The entrance of the underground parking lot is located at the downward side on the slope, in order to reduce excavation. The preservation of trees and a setback of the complex's main entrance, as well as the configuration of the plaza near the local temple shows respect towards the culture and customs of the surrounding context. In addition, a semi-outdoor activities arcade on the ground floor was created to connect with the northern and southern green belts. Last, a multi-functional gathering space placed at the location of the old community hall not only retains the historical memory of the community, but also provides a space for social interaction between old and new residents.

设计巧妙利用基地前后约 6 米的高差限制，设置停车场出入口于地势较低处，以减少土方开挖量降低环境冲击，并在停车场屋顶设计可远眺街景的休憩平台，化地形劣势为优势；考虑传统寺庙是当地居民重要的社交场所，退让住宅主入口，结合既有大树，将庙宇旁的零散地块转化为街角广场，延续在地的公共生活空间。此外，建筑底层留设半户外活动长廊，串联基地南北两端绿化带，并于临公园的南侧旧集会堂原址配置多功能集会空间，除保留在地居民历史记忆，也为新旧住民提供一处交流聚会场所。

3×3 Grid System
九宫格系统的住居平面

Due to the social trend of a low birthrate and the growth of an aging population, the project heavily features the single-room unit housing type. According to the demand for households, the "single-room unit" to "two/three-room unit" ratio is about 1:1. After careful studies, the nine-box grid system was considered to be the most efficient and suitable for the project. A symmetrical floor plan was adopted to satisfy the economic scale of the industrial production of the precast curtain wall panels used on the exterior walls.

应对少子高龄与在宅老化的社会趋势，设计中有较多的一室户住宅单元。根据房型需求比例评估，一室户与二室、三室房型总和的户数比约 1:1，精算后推定九宫格系统的住居平面为最有效率且满足项目需求条件的规划，体量配置上采用对称式布局，以满足外墙预制单元的生产经济规模。

The Construction of Precast Walls
半预制外墙工法

In order to meet the requirements of a demanding construction timeframe, and to reduce the amount of on-site finish work and material stacking, the project utilizes the semi-precast concrete construction method. This method combines a traditional poured-in-place concrete structure with on-site installation of the prefabricated wall panels, thereby providing significant savings in time and cost.

为满足缩短工期要求、减少现场装修及材料堆置不稳定之风险，项目采用半预制外墙工法，外墙板于预制厂制作、养护共计 686 片。预制外墙工厂与现场吊装同步作业，运用预制外墙板搭配现场 RC 梁柱浇置的复合工法，达到缩减工期的目标。

a Semi-outdoor arcade 半户外活动长廊
b Lobby 入口门厅
c Multiple purpose center 多功能集会堂
d Public plaza 邻里广场
e Grand terrace 阶梯景观平台
f Corner plaza 保留大树街角广场

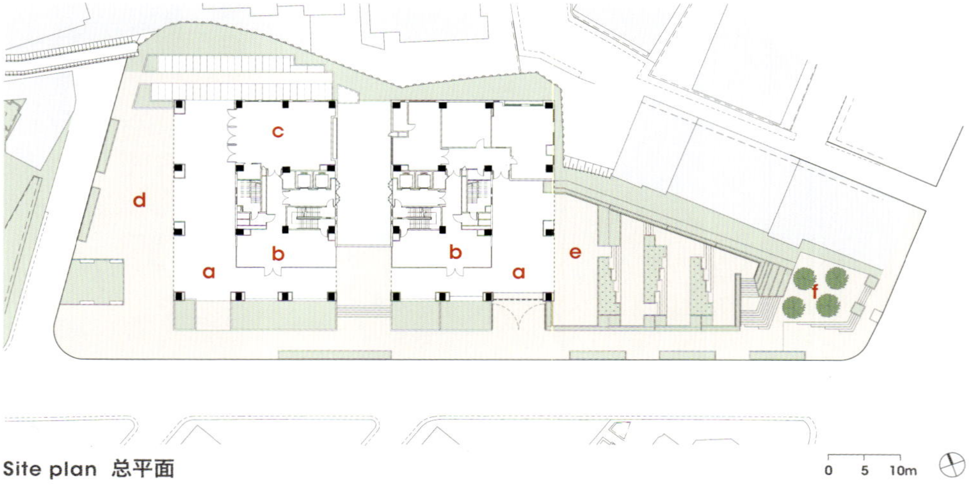

Site plan 总平面

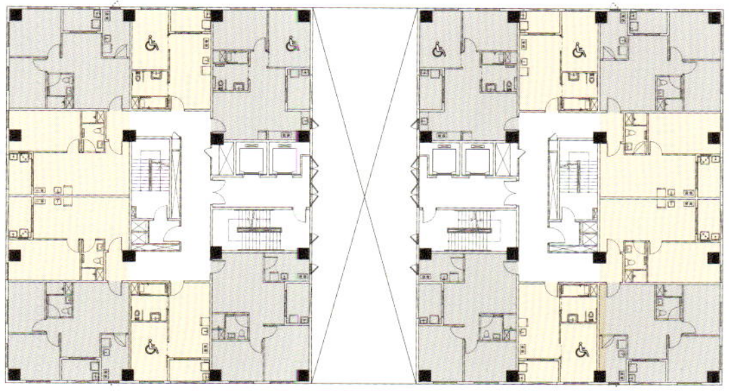

2F-12F floor plan 2–12 楼平面

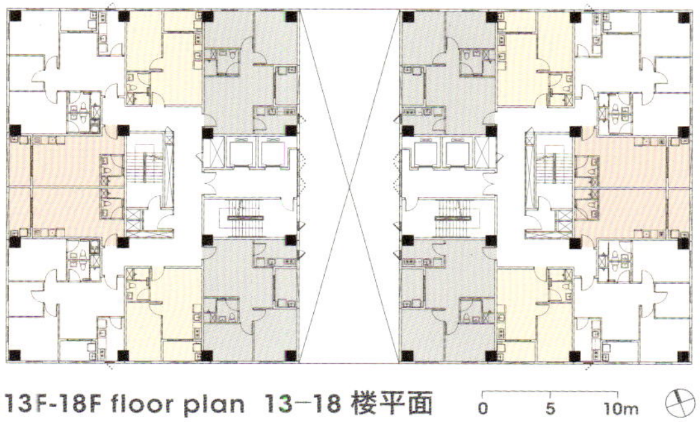

13F-18F floor plan 13–18 楼平面

- Large one bedroom unit 大单人房单元
- Small one bedroom unit 小单人房单元
- Two bedroom unit 双人房单元
- Three bedroom unit 三人房单元

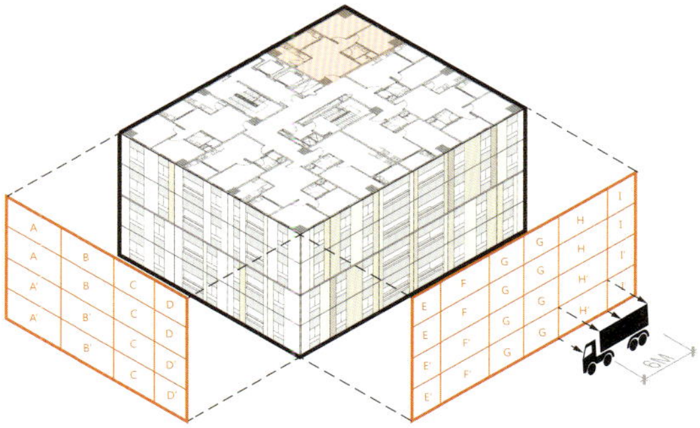

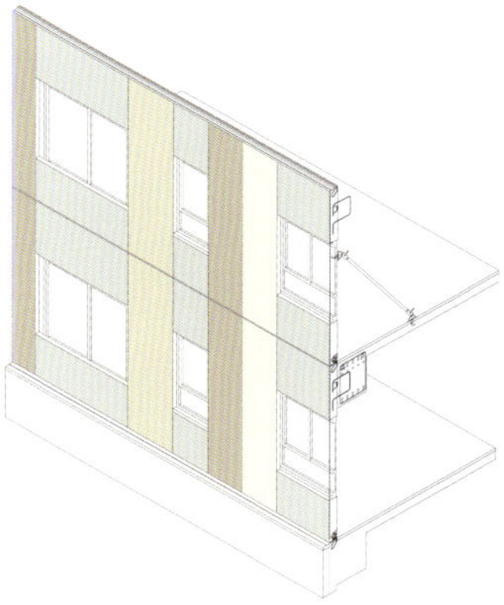

Modular prefab wall
模块化预制外墙

Ministry of Finance Joint Development Project

财政主管部门土地合作开发项目

2013–2017
Taipei
144,091 m²

State-Owned Land Activation
公有土地活化

In order to promote the policy of utilizing state-owned land and improving the efficiency of land usage, the government held a public auction to select private developers for the planning and development of the site. This project transforms what was once used only as the training institute of the Ministry of Finance into an open space for activities in the neighborhood. Through a waterscape, green fence, and plants, the previously restricted land is now metaphorically connected with the waterfront park. With old trees preserved and new trees planted, citizens can take a stroll in the green corridor, which also works to coordinate the circulation between the public and private sectors of the development. From east to west, the design consists of a commercial/residential building, a government building, the renovated financial training institute, a trainee dormitory, and residential sector. Each building has its own unique expression, all together constituting a high-quality environment for the neiborhood leading to the river.

政府为推动公有土地活化政策同时提升土地运用效益，以地上使用权公开招标遴选民间从业者进行土地使用规划及开发。设计打开旧有封闭的围墙，将原本仅供财政主管部门及所属人员教育训练的使用场所，转变为连接邻里活动的开放场域，利用公共部门释放出的公共用地串联水景、绿篱、植栽等呼应水岸公园；保护老树并新种植栽，建构一条供市民休憩漫步的绿色廊道，连接公共部门与小区动线，使其有所区分却又相互协调。设计配置由东往西，依次为面临主要道路供销售之商住大楼、财政大楼、旧建筑物更新的训练所及学员宿舍等公共部门建筑群，以及供销售的集合住宅群等。各自以其独特属性寻找最佳的呈现方式，串联后又成为一处公私双赢的互利优质环境。

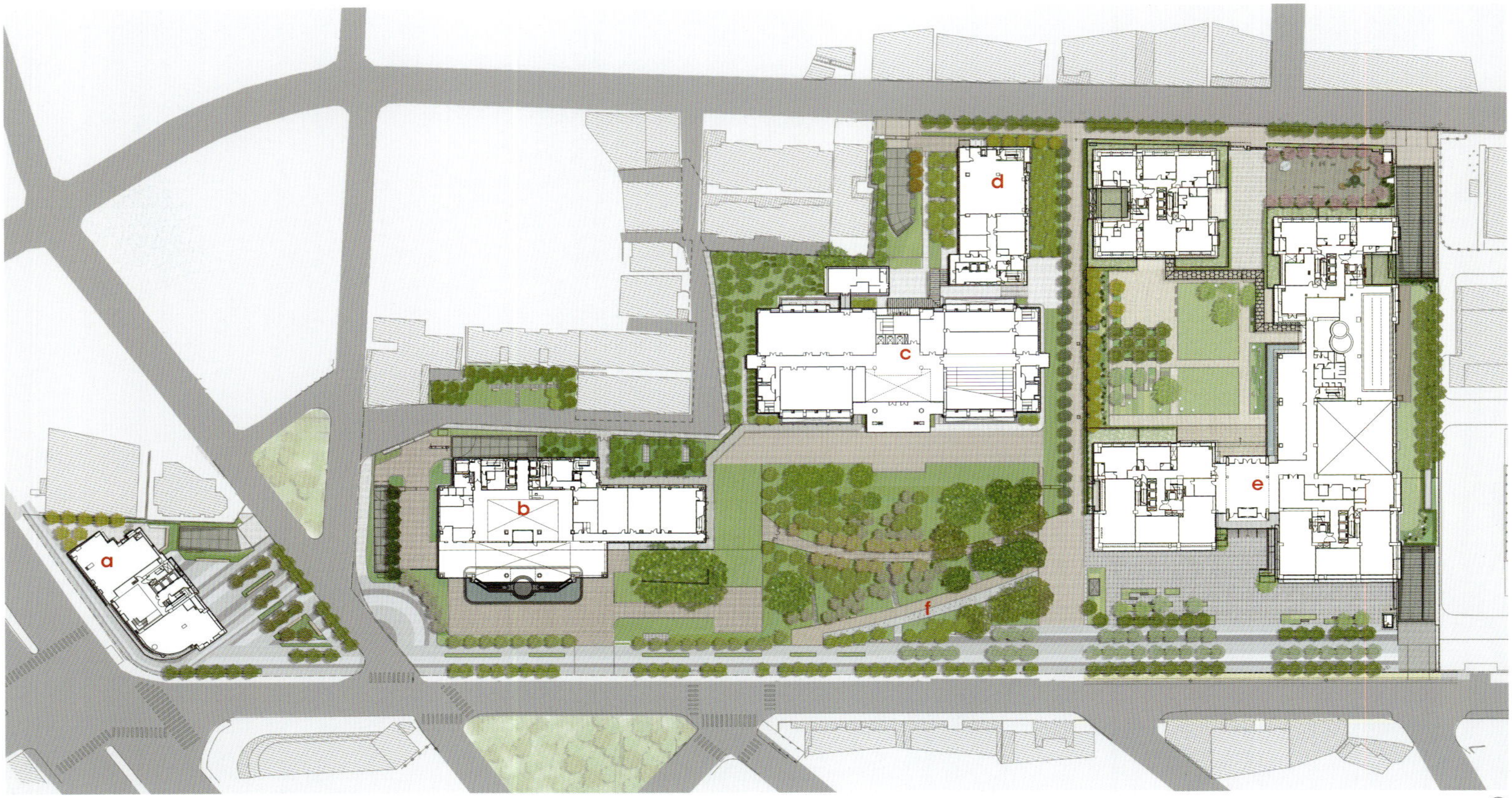

Master plan 总平面

1

a Commercial/residential building 商务住宅
b Ministry of Finance offie 财政主管部门办公楼
c Ministry of Finance training institute 财政主管部门训练所
d Ministry of Finance dormitory 财政主管部门员工宿舍
e Residential 出售住宅群
f Public park 绿化带公园

2

3-A

3-B

4

1 Public park 绿化带公园
2 Aerial view of residential landscape 住宅区景观广场鸟瞰
3 Residential landscape 住宅区中庭景观
4 Street view of residential sector 住宅群临街面

Bethany Children's Home
伯大尼少儿家园

2014–
Taipei
8,869m²

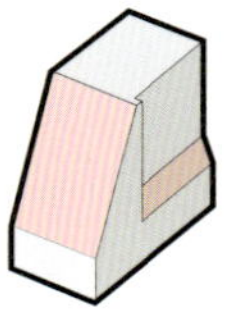

Code limitation
法规限制

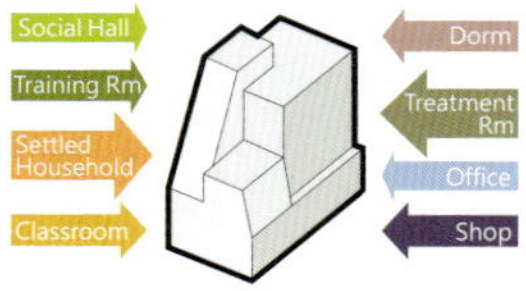

Multi-function
机能堆叠

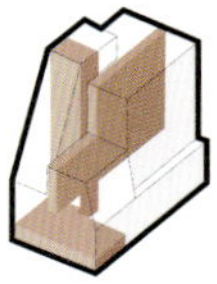

Interaction
互动交谊

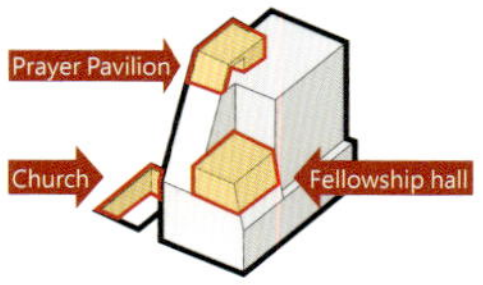

Worship
聚会空间

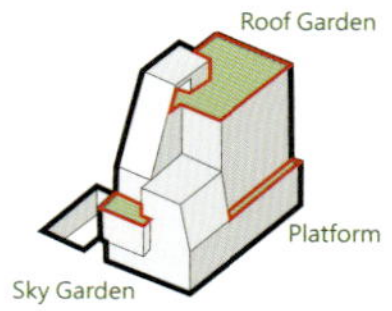

Greening
立体绿化

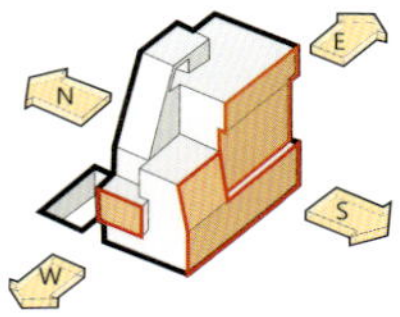

Natural light
视野采光

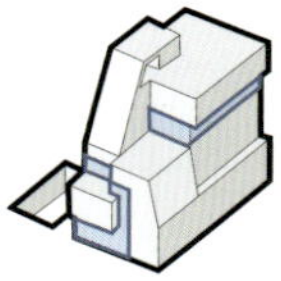

Soild & void
虚实交错

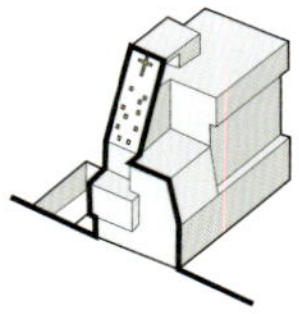

City skyline
城市天际

Welcome Home
来！回家

As a children's welfare institution, the Bethany Children's Home strives to provide a safe place for supportive care of children and juveniles from birth to the age of eighteen. This project met the needs of a disadvantaged group and created friendly connections with the community environment, while integrating the church fellowship space with the children's living arrangements and redefining the cluster relationship of the children's home. The design adopts a general family layout and room units instead of using the traditional dormitory style floor plan, enabling children and foster parents to live within a normal family life pattern.

The ground floor is an open square that provides space for activities in the community. The grand staircase and elements recalling the old church arch lead people down to the new sanctuary in the basement. The side wall of the grand staircase is covered by ceramic tile from the paintings by children, presenting the feeling of a big family highly engaged with

伯大尼兒少家園

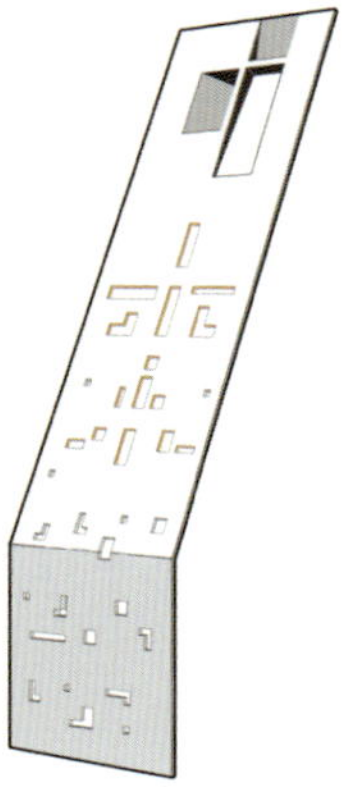

The facade represents Chinese word "come"
主墙以开口组构出中文字“来”的意象

the space. In the design of the upper floors priority is given to the lighting and ventilation, while the multifunctional activity space, staff quarters and guest rooms are distributed from north to south. The exterior design makes best use of the building setback requirements and various zoning regulations, all under Taiwan's children's resettlement act, forming a rich expression of the whole building while also effectively distributing the living functions. In addition, the main façade utilizes different openings to outline an abstract cross and the Chinese word "come," presenting the message of "Welcome to the Bethany Home"!

伯大尼少儿家园为少儿福利机构，收容扶养需专业协助的零至十八岁儿童及少年。设计重新定义了少儿家园的簇群关系，除了满足弱势关怀的需求及小区环境的友善连结外，同时整合教会大家庭的团契空间与孩童安置的生活环境。在平面规划上，有别于传统安置的“宿舍模式”，以一般住宅格局及客房单元，让孩童与“父母”拥有家庭生活模式，以便联系家人的情感。

一楼为开放式广场提供小区活动交流空间，经由景观大阶梯及带有旧教堂意象的拱门引导人们至地下一层的新教堂，景观大阶梯侧墙更镶嵌了孩童画作的磁砖，创造一个充满使用者参与感的家园；高楼层优先考虑采光通风，南北向规划分布多功能活动空间、职员宿舍及来宾客房。外观设计善用法规限制，整合少儿安置法令及土地管制规范限制，形塑建筑体量的丰富表情，并有效分布生活机能，主墙以不同的开口组构出抽象的十字架及“来”字形象，隐喻“欢迎来到伯大尼的家园！”

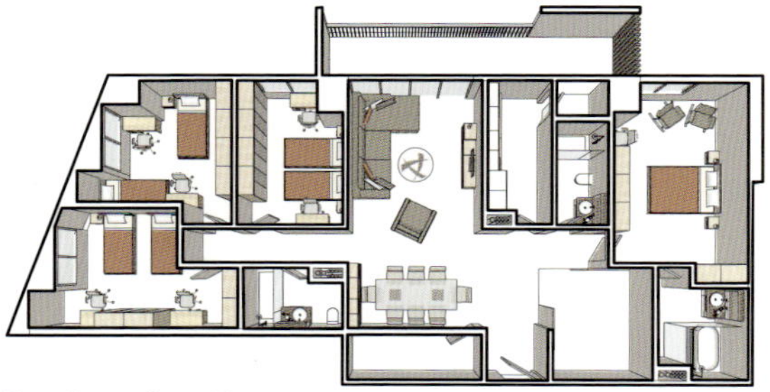

Family units with
five bedrooms and living spaces
五房两厅安置家庭单元

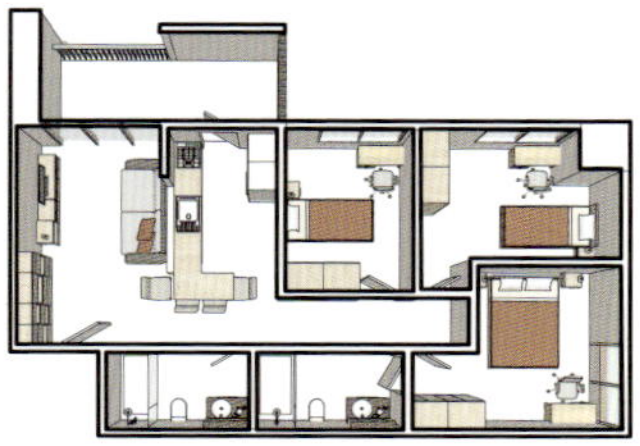

Family units with
three bedrooms and living spaces
三房两厅紧急安置家庭单元

a Fellowship hall 团契教室
b Office 办公室
c Sky garden 屋顶花园
d Training room 培训教室
e Family unit 家庭单元
f Church sunken entrance 教堂下沉广场入口
g Shops 店铺
h Church 教堂
i Carpark 停车空间

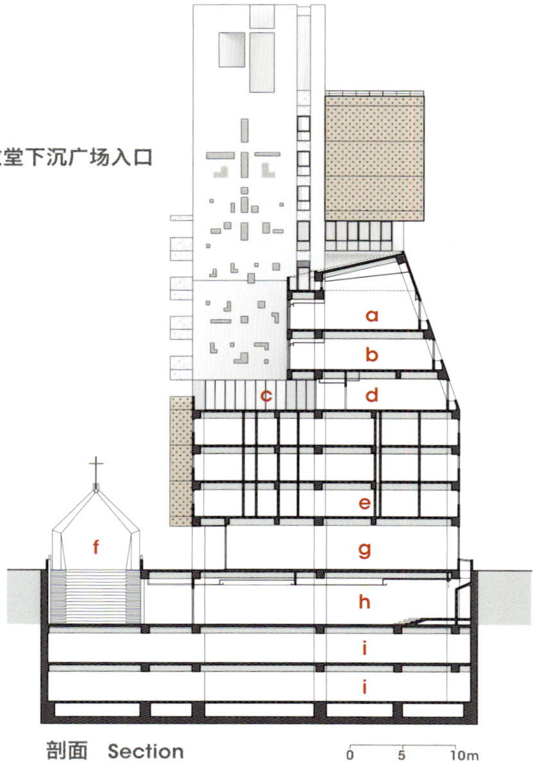

剖面 Section

伯大尼兒少家園

Campus Evangelical Fellowship Building

校园福音团契大楼

2016–
Taipei
4,381 m²

"Gospel of Kingdom will be preached in the whole world." - Matthew 24:14
"这天国的福音要传遍天下" —— 《马太福音》24:14

Rebuild the Light of the City
重建城市之光

The project came about because the original four-story building and facility have almost reached the end of their lifespan. In addition to fully integrating the original environment, the new Campus Evangelical Fellowship building also addresses the functional and spatial requirements on the limited site. Following the original operation mode, the new building consists of three units: the lower floors maintain the function of the campus library and adopts a glass curtain wall in the lobby as an expression of welcome; the middle floors are for gathering and exhibitions, and the high floors are set aside for praying, studying and office space. Given the variety of interior activities, the middle portion of the façade has a low opening ratio, which led to a window pattern that is inspired by the Braille signs of the great commission, "Gospel of the kingdom of heaven will spread throughout the world". Above the sixth floor, randomly arranged windows are adopted to reflect the diversity of activities within the fellowship, including living spaces and a rooftop garden for the students.

Existing building 原有建筑物

The cross as the main visual element is manifested at two places. One is at the corner of the sanctuary, shaping the expression of day and night with light and shadow. The other one is on the outer wall facing the main road, like a light box that attracts the attention of passersby and inspires their curiosity toward the indoor activities. As a whole, the façade directly reflects the composite functions of the interior, while also echoing the busy surrounding business district and diverse environment. It is hoped that after the completion of the project, the building will not

校園福音團契

only promote the students' evangelical work but also become a new landmark as a beacon in the city, gathering people together and offering guidance.

项目因原四层建筑空间与设备均濒临使用年限，亟待拆除重建。新大楼设计除全面整合原有环境外，亦需在此极有限的基地上，兼顾使用机能及空间效益，以创造校园团契大楼的全新面貌。延续原大楼运营模式，配置三大区块：低楼层维持原书店门市使用，以门厅挑高穿透性的玻璃帷幕表达欢迎之意；中间楼层为聚会及展示空间，顺应内聚型活动的低开窗率，将宣教大使命的内容“天国的福音要传遍天下”的点字符号，转化为开窗形式；高楼层规划为祷告、研修空间及办公室，也包括学员所需的食宿空间及空中花园，随机排列的开窗方式反映团契内部多样化的活动。

十字架主视觉聚焦于两处：其一，位于会堂内转角，以光影塑造日夜表情；其二，设置于临主干道外墙，如透光宝盒，吸引街道上人群目光，并激发行经路人对室内活动的好奇。整体而言，立面表情直接反映内部机能的高复合状态，也呼应周边热闹商圈及多元环境。期盼完工后，团契大楼不仅是推广学生福音工作的据点，更望成为散发城市之光的新地标，聚集人群，也导引方向。

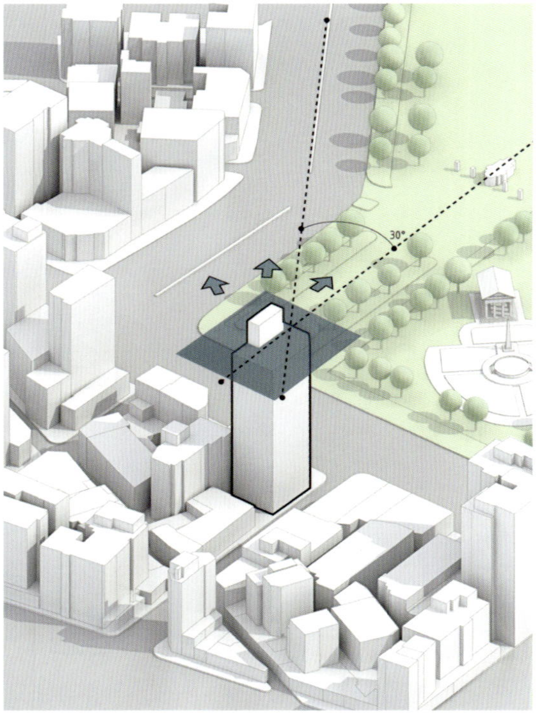

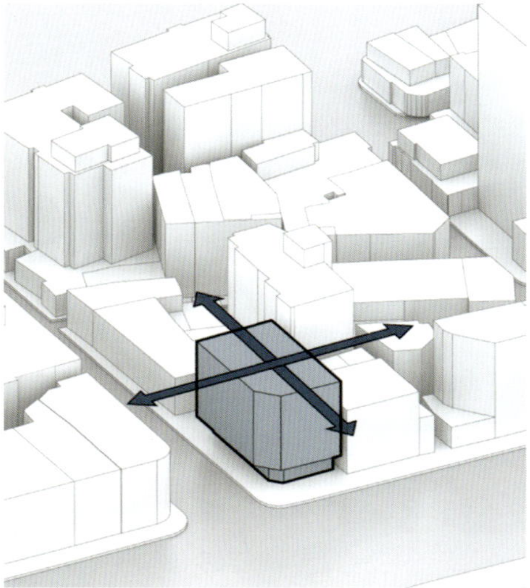

Podium mass responds to the surrounding low-rise buildings
基座体量回应周遭低层建筑物

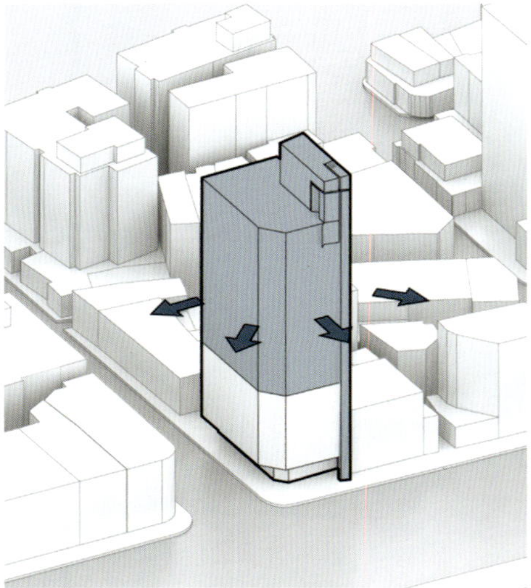

Light tower on podium mass rises above the cityscape
光之塔伫立于城市景观之上

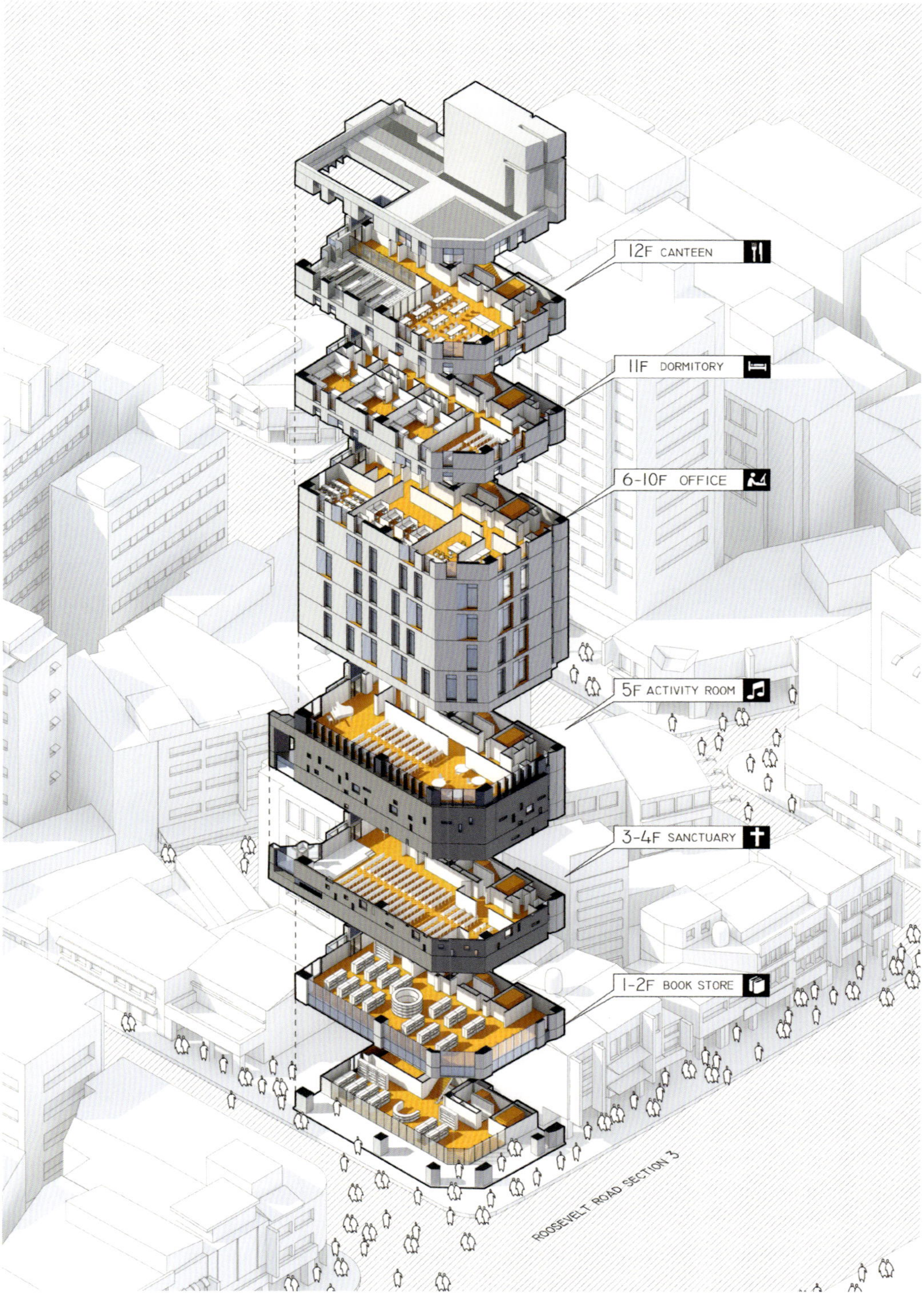
12F CANTEEN
11F DORMITORY
6-10F OFFICE
5F ACTIVITY ROOM
3-4F SANCTUARY
1-2F BOOK STORE
ROOSEVELT ROAD SECTION 3

Center for Advanced Cancer Therapies
台大医学院先进医疗基地

2008–
Taipei
146,935m²

Making a Difference in the Fight Against Cancer
抗癌医疗基地

Located on the eastern edge of the Taiwan University campus, the Cancer Center and the Radiation and Proton Therapy Center will integrate various research facilities to form the "Biomedicine Axis" within the university campus and establish an innovative hub for cancer research.

位于台湾大学东侧的癌症中心医院与辐射科学暨质子治疗中心，整合了校区医疗相关研究单位与设施，形成台大校园“生医轴线”， 完整建置了跨领域之世界级抗癌医学基地。

Empathic Design System
同理心的设计系统

The design of the Cancer Center aims to create a calming, people-oriented environment filled with natural light and wide interior spaces to provide a comfortable medical experience. The fundamental premise of the patient-centered facility is to allow for efficient service, security, and patient privacy to come together in each of the clinical and ward spaces, while introducing natural light and a landscape garden to create a natural healing atmosphere.

Housing advanced radiation equipment with a stringent requirement for safety, the Radiation and Proton Therapy Center is mainly placed below ground. The project is a compact volume with an intricate circulation arrangement. Through several setbacks of the volume, balcony gardens are created to alleviate the stress and anxiety that comes with treatment, not only for

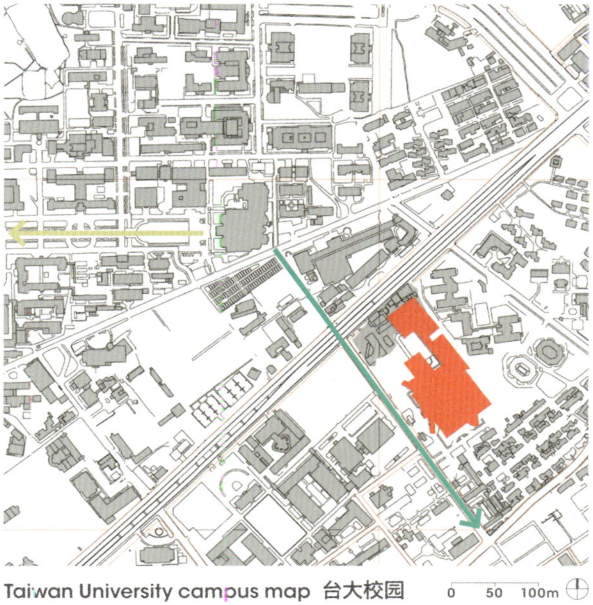

Taiwan University campus map 台大校园

Campus axis 校园轴线
Biomedical axis 生医轴线
Campus building 校园建筑
Advanced medical base
先进医疗基地

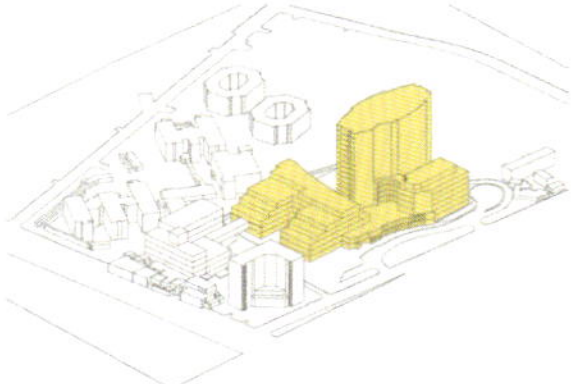

Cancer center
癌症中心医院

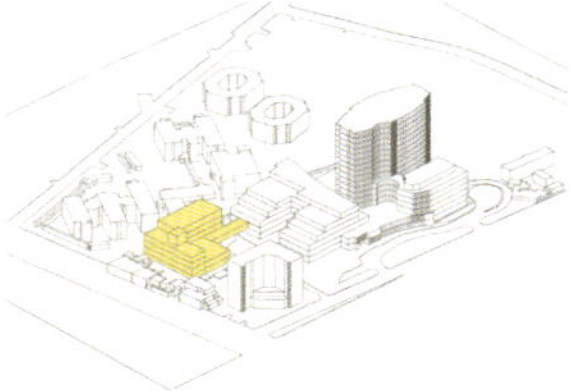

Radiation and proton therapy center
辐射科学暨质子治疗中心

the patients and their families, but also for the medical staff. Various nodes within the facilities serve as social meeting places for staff, in order to encourage interactions between the clinical research and medical services teams. Utilizing advanced BIM design, the project will be one of the most advanced radiation centers in the world and a pioneer in the field of cancer treatment.

基于以人为本的设计思考，癌症中心医院以希望光塔、翠微生活街等充满希望的环境，规划适度的自然光线与开阔的室内空间，提供有别于一般医院急促不安的就医经验；以病患为中心的医疗诊间与病房设计，满足服务效率、私密性及安全性三项要求；另引入和煦阳光与景观花园，营造有温度的自然氛围。

主体位于地下层的辐质中心，设计时即以 BIM 技术整合复杂的医疗区块及质子设备管线，配置合理化的体量及流畅的动线；利用体量退缩处塑造不同主题的空中花园，不仅降低病患与家属的焦虑感，也适时舒缓院区工作人员的紧张步调；两栋医疗设施均利用动线转折处，设置交谊角落，鼓励医疗与研究人员交流，提升研究与临床服务的互动。设计透过明亮开阔、尊重人性的空间规划结合台大专业资源，共同为抗癌医学立下新指标。

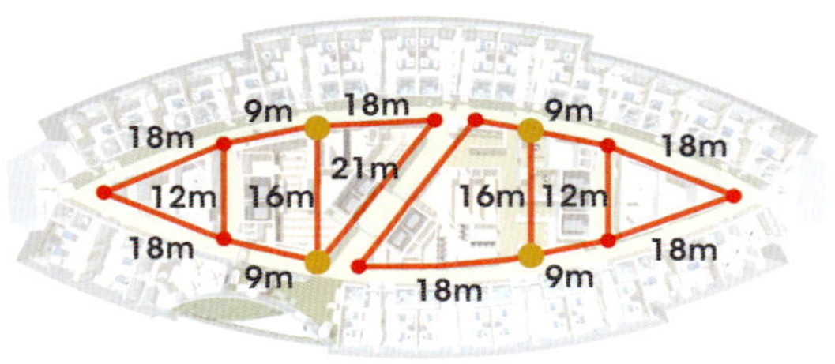

Efficient staff circulation 医护人员动线

Nurse circulation 三角形动线规划

De-centralized nursing station 分散式照护系统

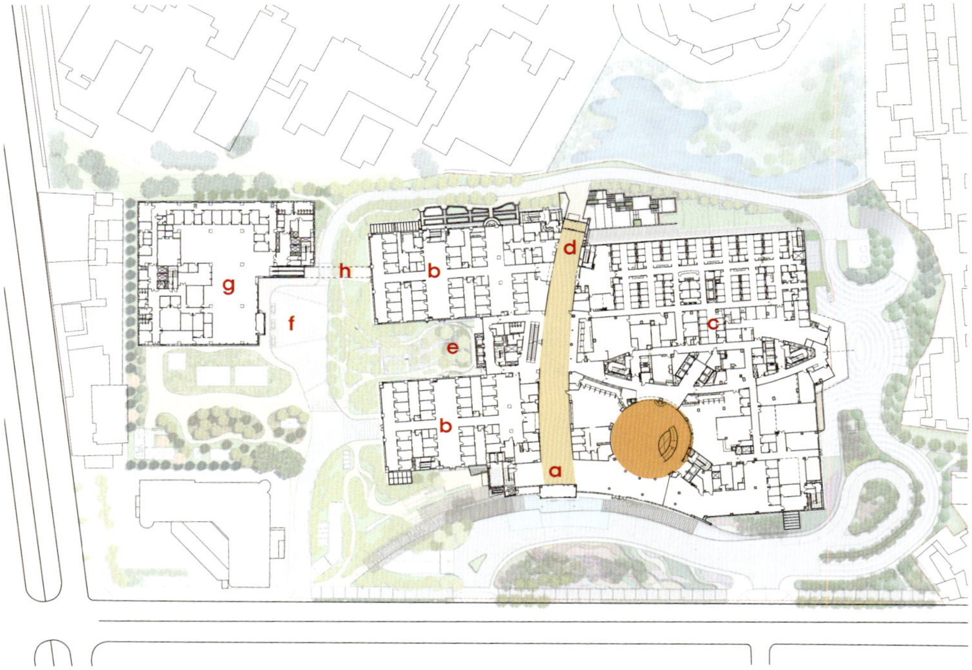

Ground floor plan 首层平面

Light tower of "Hope" 希望光塔

Green living mall 翠微生活街

- a Cancer center main entrance 癌症中心出入口
- b Outpatient department 门诊
- c Chemotherapy 化疗区
- d Secondary entrance 癌症中心医院第二出入口
- e Healing garden 疗愈花园
- f Proton center entrance 质子治疗中心入口
- g Proton center main lobby 质子治疗中心门厅
- h Connection bridge above 上方连通天桥

1

2

1 Green living mall 翠微生活街
2 Light tower of "Hope" 希望光塔

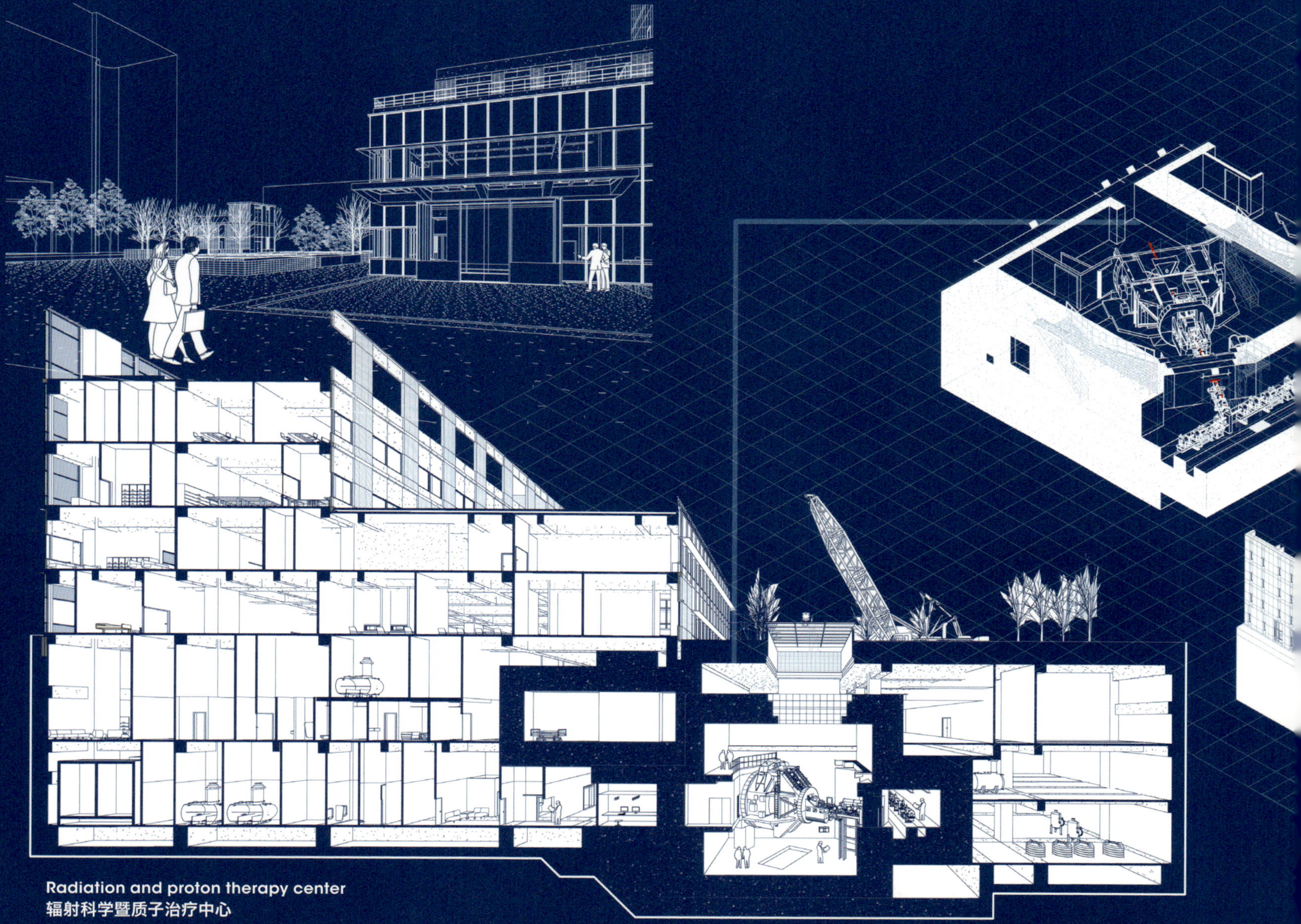

Radiation and proton therapy center
辐射科学暨质子治疗中心

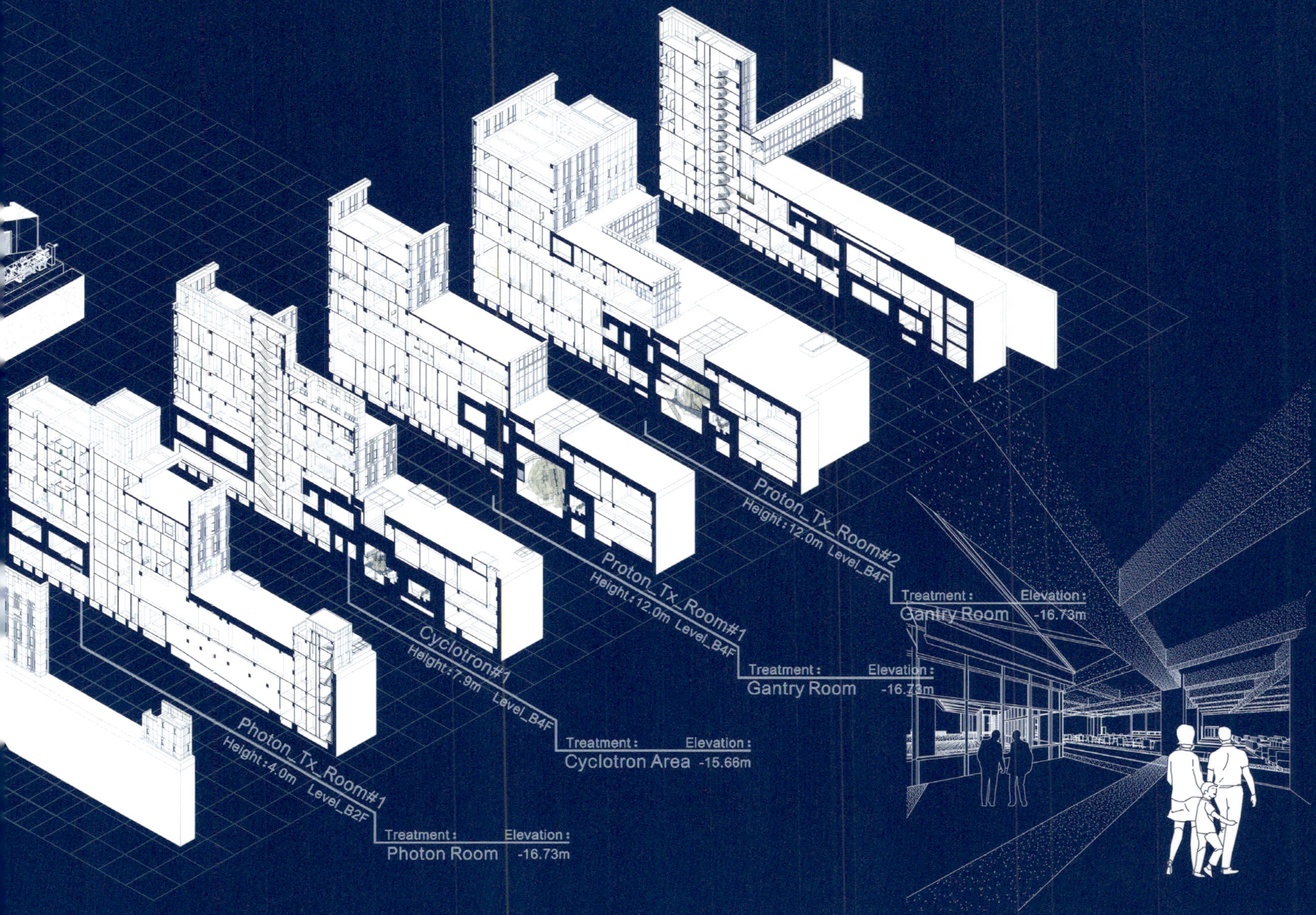
Proton_Tx_Room#2
Height : 12.0m Level_B4F
Treatment : Elevation :
Gantry Room -16.73m
Proton_Tx_Room#1
Height : 12.0m Level_B4F
Treatment : Elevation :
Gantry Room -16.73m
Cyclotron#1
Height : 7.9m Level_B4F
Treatment : Elevation :
Cyclotron Area -15.66m
Photon_Tx_Room#1
Height : 4.0m Level_B2F
Treatment : Elevation :
Photon Room -16.73m

04 SUSTAIN

Explore the essence of green architecture and build enduring partnerships in order to achieve a truly sustainable future

探究永续的本质与形式，建立互信成长的伙伴关系，共创可持续性未来

永续

ABILITY

DIA-LOGUE

WIM CHANG x CHUNGWEI SU

The Perfect Architect-Client Collaboration

张杨乾 × 苏重威

建筑师与业主的正面协作关系

Wim Chang 张杨乾

Deputy Executive Director of Delta Electronics Foundation
MSc Environment, Science & Society, Department of Geography, University College London(UCL), University of London, UK
台达电子文教基金会副执行长
英国伦敦大学地理所环境科学与社会学程理学硕士

Chungwei Su 苏重威

Principal of J.J. Pan and Partners
Master of Architecture in Urban Design, Harvard University, Cambridge, MA, U.S.A.
潘冀联合建筑师事务所主持人
美国哈佛大学建筑暨都市设计硕士

Time/Location: Summer, 2017, Delta Electronics Foundation, Taipei
对谈时间 / 地点：2017 年夏，台北台达电子文教基金会

SU: Our first project together was Phase II of Delta's Tainan Plant back in 2010. Well, I guess we did something right, so our collaboration continued on afterwards, from the Ring of Celestial Bliss in 2013, School Heritage Building in 2014 to the Delta Americas HQ in 2015. In addition, Mr. Pan has been invited to many events such as the COP21 in Paris in 2015, and the Delta Solar Building Competition and the Peking Qinghua University Green Building Exhibition in 2016.

CHANG: I recall that after we started working together, I became really curious about JJP as a firm. So I read Mr. Pan's autobiography, *Brick by Brick – Building the Foundation of Life*. Then I realized that Mr. Pan is a real positive and productive force in architecture. In this regard, both of the founders, Mr. Bruce Cheng and Mr. Pan, are very much alike.

SU: A corporation's culture is cultivated by its founder, so I would like to talk about them first. Five years ago, we realized Mr. Cheng's insistence of "green energy" is similar to Mr. Pan's dedication to "green building". In the early years of the firm, Mr. Pan did not have the modern green building design tools when he was working on the ITRI Center for Measurement Standards. He could only rely on his comprehension of the environment to guide his design, such as having the fenestration respond to its orientation. The Chung-Yuan Christian University Main Library is the first open-access library whose design realized the principles of "green building". Since there was no central air-conditioning, the building was designed to facilitate airflow as well as thoughtful fenestration to accommodate thermal energy and lighting.
On the other hand, the government, who has constantly been promoting sustainability, does not consider it from a fundamental point of view, but as an add-on after all the essential decisions of the project are made. However, I believe both firms approach sustainability from beginning to end. As a result, this shared belief has made us great partners.

CHANG: Back in 2012, as we were preparing for the Taiwan Lantern Festival, performing a building carbon footprint audit for the project became a requirement. We purposed two main points: 1. All materials should be recyclable, 2. Carbon footprint data should be collected for evaluation. Furthermore, post-festival carbon footprint data should be added into the building life cycle calculations. The result was quite significant - the total carbon emission was only 20% of a comparable reinforced concrete building. The project was also the first "building" to officially complete the carbon footprint audit in Taiwan. In the end, the total amount of carbon emissions produced in the 15-day Delta Ring of Celestial Bliss exhibition is 94.7 tons, which is only 22% of the total carbon

"In the case of JJP, Delta just proposes some basic ideas. What really matters are the subsequent communications with the architects and the construction company."

"但以我们合作这几个项目来说，基金会提的是粗浅的想法，真的要落实需要跟建筑师甚至施工方共同讨论。"

苏：JJP 与台达电及文教基金会合作的时间，大约是 2010 年台达台南二期厂房开始，应该合作愉快（笑），所以接续一连串的合作项目，从 2013 年元宵灯会的永续之环、2014 年台中一中校史馆修复、2015 年台达美洲总部，2015 年潘先生多次受台达电之邀参与在巴黎举行的联合国气候变迁高峰会，一路到 2016 年台达电太阳能建筑竞赛新闻发布会，以及北京清华大学"绿筑迹建筑展"等建筑委托项目和多项活动参与。

张：我记得基金会跟潘冀事务所的合作之初，那时候就对事务所蛮好奇的，直到阅读了潘冀建筑师的《人生基本功》之后，觉得潘建筑师是建筑界的实在力量，也觉得两位创办人郑崇华先生与潘冀建筑师非常相像。

苏：企业文化都是由创办人建立的，当然先从两位创办人谈起。回想五年前与基金会讨论"永续之环"时，当时就发现郑先生对于"绿能"的坚持，而 JJP 的"绿建"精神也是承自潘先生创办事务所以来的理念。早年潘先生设计工研院的精密仪器中心，还没有太多工具及技术去切入"绿能"运算，只能凭着对环境的理解与认知设计，针对四个不同面向，设计出各自对应的开窗；1980 年代的中原图书馆是台湾第一所开放式的开架图书馆，在当年没有中央空调的情况下，延续开架的空间精神让内部空气流通，再透过不同朝向的开窗形式采光，调节热能及光源，实践"绿色建筑"。反观公共部门一直推动的可持续项目，大多不是从根本开始考虑，而是等所有设计立面都决定了之后，才开始思考如何与可持续产生关联。所以我觉得我们这两个企业对可持续的看法相近，都是从最基础做起，从源头看起，所有绿色建筑做法一定是自概念开始就一路发展到各运行时间直到完成，"从根本做起"的精神是非常相似的，也显现在我们一起合作的项目上。

emissions produced in the annual Taipei 101 fireworks show (430 tons).

When Delta first purposed these "stringent" requests, it must have been be a real "pain in the neck" for JJP, I guess.(smile) Especially considering there were only five months from the kick-off meeting to opening. The schedule and budgets were both extraordinarily tight.

SU: Actually, JJP has always believed that projects with limits or certain rigorous conditions would only make the project better, because those limits are the true responses to the environment and site conditions. If the client has a clear objective that's difficult to attain, it would motivate us to innovate harder and develop an even better solution.

Taking RCB as an example, when it was still in the conceptual design stage, our team asked for input from the Delta projection team. Specifically, on specifications like screen distance, quantity and positions of projectors, projecting on a curvilinear

张：基金会与事务所在 2012 年筹备来年的台湾灯会时，希望将建筑碳足迹稽核认证重点导入，所以，找合作建筑师时，即提出两点：一、希望建材能够全数回收再利用；二、希望能提供细部数据，以供建筑碳足迹评估系统计算使用。最后也确实将建筑生命周期完全纳入碳排量计算，成果比一般混凝土建筑减少近八成碳排放量，也是台湾第一座经过检验并对外公告"碳足迹"的建筑物。最后成果令人满意：15 天展期的总碳排放量（94.7 吨），仅为 101 跨年烟火一个晚上总排碳量（430 吨）的 22%。

但当初基金会在提这些要求，对事务所应该也是一大难题吧（笑）？特别是从决定要执行到要开展，10 月决定，2 月开展，仅 5 个月，其实没有太多时间可以讨论，还有预算需要考虑。

苏：其实，我们一直相信设计上的条件或限制，反而是把设计做好的原因，因为这些限制才是响应不同基地所在之地景与环境的基础条件，响应人性及需求才能成就人本设计。如果业主有很清楚的目标，即便目标看似难以达成，但它却是帮助我们突破与创新设计的动力，成为我们共创的成果。

以"永续之环"来说，早在设计构思之初，就希望能结合台达电的投影技术，但在设计上需要符合如屏幕距离、投影机数量及位置、环状投影与渐变形状的屏幕等专业条件，需要仰赖台达电的专业一起成就；至于如何将投影幕做成一个有表情的建筑，在设计的过程当中，除了技术外，我们更希望能透过建筑的形体与围塑的空间，传递大家对于环境关怀的共同信念。

整个合作过程，真的是因为双方都有一致的共识，才能达成所有资源百分之百可再利用的目标！

张：记得当时提了很多种再利用方案，以建筑主体的钢结构为例，从钢结构拆卸、存放到重组，花费了许多心力讨论与设计，虽然为了取得建造许可证多花了些时间，但终究在 2015 年底运往台东县太麻里乡的大溪小学组装，设计上也针对台东县在台风季节所需面对的严峻气候条件作了修正，即使历经 2016 年超强台风的肆虐，兼具美感与实用性的风雨球场仍屹立不摇。

苏：除此之外，我自己觉得有意义的是，我们去灯会现场看，经过其他的灯区，只见原本公园草地都打上了水泥跟柏油，只有我们"永续之环"因为地坪全数使用天然材料，因此活动结束后无需进行地坪铲除作业。光是选择铺面材料衍生的

surface etc. With Delta's key technical assistance, we were able to shape a spatial volume that represented our faith of respect towards the environment. It is because of the close collaboration between Delta and JJP, that RCB could reach its 100% recyclable material target

CHANG: I remember there were a lot of proposals about how RCB could be recycled and reused. Taking the steel structure for instance, the process of dismantling, storing and reassembling took a lot of effort to discuss and design. Though the building permit took a lot of time, the steel structure was finally delivered to Taitung Daxi Elementary School in 2015 as a weather canopy for its outdoor basketball court. Necessary adjustments were made to adapt to the weather conditions of Taitung and the structure survived Typhoon Nepartak in 2016.

SU: In addition, what I found meaningful was that other lantern areas had concrete or asphalt paving, while RCB utilized natural materials such as bamboo, so there was no extra demolition work afterwards. I could not help but wonder why didn't the other lantern areas adopt such a simple but significant method to protect the environment. I believe we achieved a different result because Delta requested all materials should be natural at the onset. Both Delta and JJP hoped to convey that it is possible for such holiday events to only leave a small environmental footprint. However, it is a real pity that I haven't seen anything similar since then.

CHANG: Some expos have started applying "carbon neutralization" concepts but not in any significant depth nor scale.

COP 21 Paris, 2015 巴黎联合国气候变化大会

效应就非常明显，让人不禁会想为什么其他展区或灯饰不能也这样应用简单、好看又环保的材料呢？我相信是因为自始就有着“所有建材都要是自然”的出发点，才会有不一样的结果。
透过这个合作我们最希望收获的效应是让更多人知道：办这类的节庆活动，是可以做到不需要对环境造成过大的冲击的，但是往后这几年灯节，始终没有再看到相似的设计做法，有点可惜。

张： 有一些博览会开始思考以碳中和的方式去执行，但的确没有大规模的试做推行。

苏： 当然没有人来模仿设计是好的（偷笑），但也算是不好，毕竟，如果“永续之环”能在传统节庆的台湾灯节带来低碳思潮，进而影响社会推动绿能、迈向可持续，才是我们真心希望的……

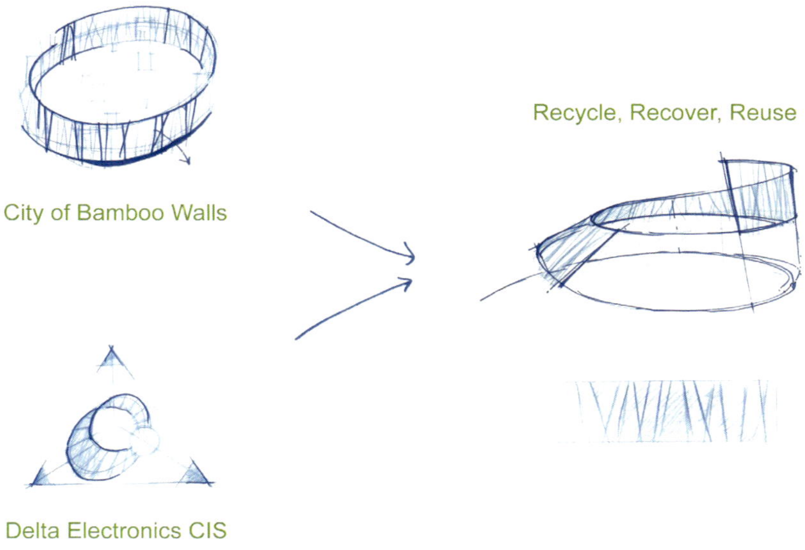

Ring of Celestial Bliss 永续之环

Sports court shelter of Daxi Elementary School 大溪小学风雨球场

SU: Of course it's a good thing that our design hasn't been imitated, Or maybe not. After all, we hoped RCB could help introduce the concept of low carbon footprint emission and influence the society on a path towards greater sustainability.

CHANG: Another project we worked on together was the School Heritage Building of the Taichung First Senior High School (TFSHS) . I think members of the Delta Foundation, the school faculty, students and alumni were all amazed when they saw the renovated building.

The building was severely wind-erosive and not well ventilated. A tin-roof was added at some point to mitigate a long-term water leakage issue but it also heated up the indoor temperature and made the auditorium like an oven. Honestly it was hard to imagine such a dilapidated historical building could be reborn. The original objective was to have the building achieve negative carbon emissions, which was difficult because the building was over 76-years old. This is a

张：另外一个合作项目，台中一中校史馆，我相信不论是基金会或台中一中的师生校友们，都对于以旧复旧的结果感到相当惊艳。原本的校史馆状况恶劣，由于墙体风化严重、室内通风不良，加上原始屋顶因为长期漏水改铺红色铁皮屋顶而造成增温，让整间礼堂如烤箱，在修复之前很难想象可以运用绿色建筑方式赋予历史建筑物新的生命。

最初是希望它能达到负排放，但因为这座建筑已有 76 年的历史，比一般建筑五六十年的使用寿命要长；此外估算建筑修复后的用电，应可以被基金会在旁边敬业楼顶新增的太阳能板发电中和掉，甚至是挑战负排放的目标。

后来虽然来一场大台风，对新设的太阳能设施造成些伤害，但现已全数修复发电，让校史馆更具时代意义。虽然这栋建筑物不大 (建筑面积

really advanced age given that most buildings only last 50-60 years. Furthermore, it remained a question whether the building's electrical usage could be offset by the solar panels that Delta installed on the adjacent Jing Ye Building. Such were the challenges towards the goal of negative carbon emission.
Though a strong typhoon had caused some damages, the solar system is now back online, making the School Heritage Building even more significant. The footprint of the building is not large (coverage: 656.68m²) but we still felt JJP devoted a lot of effort to this project. JJP patiently cooperated with the Delta Foundation and TFSHS to navigate through the government's red tape. Mr. Pan also came to the site for inspection several times even monitored the model construction himself. We felt JJP really took good care of Delta.

SU: Given the historical value of the building and its special meaning to Mr. Cheng, we were determined to complete this project to the best of our abilities. The building had been listed as a cultural asset by the government, so we really needed to collaborate with a historical preservation team. From the design aspect, we wanted TFSHS to be able to use the building based on its current needs. We did not plan to simply restore the building since it would be meaningless if the building was just a historical specimen after renovation.
With that said, we kept the original walls while attempting to restore the historical atmosphere of a Japanese-style auditorium. We also considered the intended new use of the building and introduced some green building design concepts. Since occasional gatherings will take place on the ground floor, we add a lightly constructed mezzanine to accommodate the school history exhibition. The key to a successful transformation of a historical building is to achieve the perfect balance between the old and new elements. They should complement and not outshine each other.
Furthermore, we also replanted trees around the building to enhance shading. Based on the current usage of the open spaces, I think there should've been even more trees. From an architect's perspective, it is easier to take good pictures without trees that may cover the building, but having some trees around the building certainly benefits the surrounding environment.
Although the project scale was indeed small, the challenges were relatively huge. If we hadn't thought through every detail at the beginning, the resulting imbalance of the architectural elements would forever destroy the building's original spirit and heritage. We must do our best so we can properly preserve the heritage and breathe new life into the building.

CHANG: Actually the historical research information used on the renovation is also great learning material for the foundation. And because of JJP, we were able to observe how the construction company was able to renovate without any impact on the current condition. Since the brick tiles

656.68 平方米)，但过程中却能感受到事务所对这个项目的全心投入，不但耐心地陪着基金会与学校完成漫长的文物资产审查，潘先生也好几次到工地作现场勘验，甚至后来制作模型时，也是亲自到场盯着制作细节，让我们觉得事务所真的很照顾台达电。

苏： 因为这栋建筑的历史价值以及对郑先生的特殊意义，当初我们就觉得一定要协助做好这个项目，又因为其中涉及的历史建筑物修复与文物资产保存，也积极找寻合适的文物资产保存团队一起参与，包含后段的修复，实际动手的修复工作都是由专业熟手亲自操刀；从设计来说，我们希望修复后，学校能以符合当代需求的方式继续使用这栋建筑物，而不是以恢复当初建筑百分之百原样的方式规划，如果修复后只是成为建筑标本，

Taichung First High School Heritage Building 台中一中校史馆修复及再利用

Delta Tainan Plant Phase II 台达台南二期厂房

are no longer manufactured in Taiwan, the contractor even hired two very seasoned masters to remove the damaged ones and replace them with tiles from less visible locations of the building. The work of the tile was extremely sophisticated. If we did not collaborate with JJP, we would not have known the details.

SU: In regard to sustainability, we also have learned a lot from the foundation. For example, some minor differences in design that seemed to have no major impacts on the environment actually turned out to be significant. When designing Phase II of Delta's Tainan plant, we were asked to use translucent glass as partition between hallways and washrooms in order to bring in more daylight thereby saving energy. We also learned a few architectural design techniques that help the building equipment operate more efficiently. These decisions were simple but effective. We learned these lessons from Delta and Delta was generous enough to let us apply them on other projects.

意义就不大了。

所以，除了保留属于原有历史建筑物墙面部分外，也试图呈现日据时代中学礼堂的历史意义，并纳入现代高中校园使用考虑，添加绿色建筑思维的方向思考。为满足校史馆以及不定期集会空间的需求所设计的轻巧夹层，可在不影响地面层完整空间灵活的利用情况下，提供展示的功能。

此外，我们觉得历史建筑再利用成功的关键，是要理出原有建筑与新建部分的“平衡关系”，并保持和谐不互相竞争，主从关系不致过度消长的状态。

此外，也试着把原本在校史馆旁的树种回去增加遮阴效果。只是我觉得依学校现在的使用模式跟开放空间的功能，树种得还是少了些，有些可惜。虽然从建筑师的角度来看，拍照时没有大树遮住建筑物取景可以容易些，但房子旁边如果多一些树，对空气质量与环境都会更好。

校史馆确实是一个规模很小的项目，但反而是因为小显得挑战更大，如果没规划好的话，可能会因为主从关系失衡而破坏了原本历史建筑的精神，因此失去见证历史场所的机会，过程中自然就更加自我敦促必须尽力完成，才能让古迹新生。

张：其实事务所提供的史迹修复信息，对于基金会也都是学习素材。也因为事务所的关系，我们才有机会看到承包的施工方如何以旧复旧，甚至愿意找两个人瑞级的老师傅，花上两个月时间，把位置不明显的旧磁砖敲掉，贴到另外一边有破损磁砖的扎眼位置，因为台湾已经没有生产这样的磁砖，做工非常细致。如果没有跟事务所合作，基金会真的学不到这些细节。

苏：在可持续这方面，我们从基金会也学习到很多，像我们原本觉得微乎其微的设计差异性，似乎对环境没有太大的影响，后来才知其实不然：比如我们在设计台达台南二期厂房的时候，当时就要求所有的厕所跟走廊，必须使用透光的玻璃隔间，来提供自然照明，减低照明能源；还有一些是设备方面的节能手法，透过建筑设计的合理化降低对环境的负担，这些决定很简单、普通但有效，都是我们在与台达电的合作中学来的，而这些经验你们也很大方让我们应用在其他的设计上。

张：站在基金会的立场，多用肯定是好事！

苏： 基金会其实不只跟 JJP，还有跟其他的事务所合作，JJP 是否有别于其他事务所不一样的地方？

CHANG: From the Foundation's point of view, it is never a bad thing to have more sustainable designs.

SU: Since the Delta Foundation also works with other firms, how is JJP different?

CHANG: Actually each firm has its unique culture. The foundation runs a "Green Collar Workshop", where we invite architects that specialize in different fields such as water resource management, application of solar energy, and passive design. We have adopted different ways to work with each architect. In the case of JJP, Delta just proposes some basic ideas. What really matters are the subsequent communications with the architects and the construction company. In addition, JJP produces fine work efficiently at a reasonable price (laugh). The important point is we collaborate on the projects together.

When we planned to construct the Ring of Celestial Bliss, we truly had the intention to reuse all the building materials but no means to do so. While there are countless ways to recycle or reuse building materials, it is much more difficult to achieve the expected target if we don't start in the design phase. We could not have achieved the result with Ring of Celestial Bliss without JJP's assistance. I think that's how we see JJP is different from others. JJP could take the imaginative ideas, including Mr. Cheng's, and turn them into reality. This is what we truly respect and admire.

I recall that Mr. Cheng once remarked to Mr. Pan that it's pity they could not have met earlier. So I think Delta and JJP are both learning from each other while accomplishing some dreams together. We truly have learned a lot from JJP, especially in terms of techniques. I think how far can the concept of green building go depends on architects like JJP. A net zero building can only be achieved by walking the walk, not just talking the talk.

SU: We believe that we need to make every design as environmentally friendly as possible. Although not every client truly supports this idea but we are certain that Delta firmly believes in this idea and will always act sustainably. Only environmentally friendly buildings can be sustainable. On an even higher level, we hold this spirit when we work with our clients. By working respectfully on the projects, the clients will often understand and support us. In the long term, a well-developed and supportive working relationship leads to a greater outcome for both parties.

We hope we will have more opportunities to work with Delta, not only on sustainable buildings, but on a sustainable business relationship.

张： 其实每个事务所特性都不一样，包括我们自己举办的“绿领建筑师”工作坊，请来的都是不同事务所的建筑师，有的专长在水资源管理，有的在太阳能再生能源运用，有的是被动式设计，不同的合作对象就有不同的学习方式。但以我们合作这几个项目来说，基金会提的是粗浅的想法，真的要落实需要跟建筑师甚至施工方共同讨论，而且事务所速度快，价格又很合理，质量还很好（笑），重点是可以共同发展完成这些任务。

老实说在“永续之环”那个项目的时候，我们虽然希望建材可以全数再利用，但一直都没有把握。讲当然比较容易，我们可以邀请不同的国际大师来告诉你建材可以如何再利用，但这作品若不从设计阶段就去构思，是很难达到预期中的成效的。老实说若没有事务所的帮助，基金会徒有高空的理念，完全没办法落地，这点倒是我们看到贵所跟其他事务所不一样的地方。我们所提的这些天马行空的概念，甚至郑先生的一些想象，事务所都能达成，这是让我们相当敬佩的。

我记得有一次郑先生跟潘先生共同出席一个场合，郑先生对潘先生说相见恨晚，可以共同去成就一些梦想，我想彼此也都是在学习，尤其是在技术方面，让我们在绿色建筑的概念上多走了几步，但真的要能实践，还是要靠事务所，不是光用说的就能达成净零耗能。

苏： 我们都认为只要有机会，就应该把设计做到对环境最好，虽然不见得所有业主都会真心支持，但是我们很确定台达电是真正相信，而且会如实地往可持续的方向去执行的。建筑要做到对环境友善才能永续；扩而广之来看，我们把这样的精神一并用在共同工作的业主伙伴关系上，当以友善尊重的态度看待所有经手的设计时，经常可以获得业主相同的理解与支持，长期下来这种健全良好的互相支持成长互动模式，也让共同完成的设计成果得到指数级的效益。

我们相信与台达的可持续合作能在更多元的平台上持续发展，不仅成就建筑的可持续，也在“合作永续”上造就典范。

Architectural Terroir: Delta Electronics Series
建筑接地气：台达电子合作系列

1

Ring of Celestial Bliss
永续之环

2012–2013 Ø 30m

2

Delta Tainan Plant Phase II
台南二期厂房

2010–2012 27,487m²

3

Delta
Americas Headquarters
美洲总部

2011–2015　　16,648 m²

4

Taichung First High School
Heritage Building
台中一中校史馆修复及再利用

2012–2015　　822m²

To achieve "sustainability" not only requires design thinking on the relationship between humans and the environment on a fundamental level, but also the support of like-minded partners. From the cradle-to-cradle "Ring of Celestial Bliss", passive designs on the Delta Tainan plant expansion, the net zero energy target of the Delta Americas headquarters, and the restoration of the School Heritage Building at Taichung First High School, these four projects would not have been possible without the collaboration between the architect and clients. The fruits of our collective efforts demonstrate an enduring humanistic concern for the environment and its precious resources.

“永续”价值的实践，需要可持续性的设计思维，从根本响应人类生活与地球生态环境，也需要理念相同的伙伴共同推动。从摇篮到摇篮的“永续之环”、被动式设计的台南二期厂房，透过建筑设计与科技达成净零耗能的美洲总部，到绿化历史建筑物的校史馆修复等，透过建筑师与业主共同对环境与人性关怀发展出的四件作品，持续在不同场域发挥影响力，提醒着人们珍惜地球资源。

Ring of Celestial Bliss
永续之环

2012–2013
Hsinchu
Φ30m

Complete Life Cycle Thinking
全生命周期的绿色思维

As the main lantern for the 2013 Taiwan Lantern Festival, the project is a pioneer in carbon footprint documentation for its type. The project also uses 21.3% less energy than a comparable building. The design conveys a profound respect for nature while providing an auspicious vision for the future. The choice of form and materials used for the lantern is inspired by the historical and cultural characteristics of Hsinchu, whose ancient name was the "City of Bamboo Walls." The locally sourced construction materials have the additional advantage of being environment-friendly with a reduced carbon emissions footprint. From the outside, the lantern appears as a glowing object hovering in the night, a feat achieved through the special and innovative design of the steel structure. Inside the lantern, a ring of ever-moving images produced by the latest projection technology and LED lighting serves as a metaphor for nature's endless cycle of life and inspire hope for the future.

“永续之环”是为台湾年度节庆——元宵节所设计的临时建筑，为全球最大的节庆灯体，也是少见的实施全生命周期碳排放稽核记录的装置，碳足迹仅为同样规模建筑的 21.3%。设计之初，即建构完整的材料再生机制，希望民众在赏灯之余，也能对环境有所反思。整体造型以生生不息的生命之环诠释中国人“鸿福齐天”的美好愿景，外环立面以竹材回应新竹古称“竹堑”的地域特色，透过钢结构的特殊设计，让前倾后仰的渐变结构，悬挑漂浮于夜空之中，内环以回收材料制成的投影布幕，结合 270° 投影技术环场动画及全彩光雕，全部材料皆可拆解并回收再利用，充分展现设计新意与可持续性价值。

Cradle to Cradle
从摇篮到摇篮

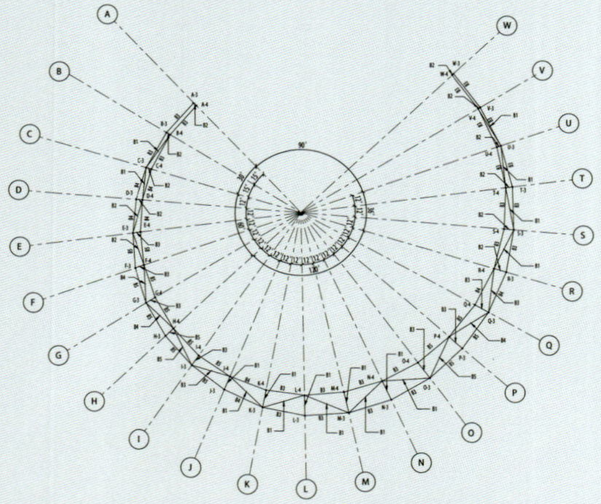

Structure layout 结构布局

Cradle-to-cradle design:

1. **Screen:** Through the great skills of the factory workers employed by a special education center, the large fireproof projection screen has turned into hundreds of wonderfully designed environmental friendly tote bags, which are used for energy educational courses in schools.
2. **LED lighting:** 234 sets of LED lighting fixtures were donated to the Taiwan Museum of Natural Science to illuminate the "Tropical Rainforest Greenhouse" exhibition. The light show uses the theme of "Magical Rainforest – Four Season Symphony" to raise awareness of the rainforest crisis.
3. **Bamboo wall:** The 500 pieces of Makino bamboos installed in the perimeter of the lantern were donated to Earth Passengers, a non-profit organization, which used the materials to build an education workstation equipped with photovoltaic panels and rainwater collectors.
4. **Paving:** The bamboo paving was left in place to decompose to provide nutrients for the park vegetation.
5. **Projection booth:** The nautilus-shaped projection booth originally housed several projectors and control equipment. The structure is now a green architecture DIY experiment cabin for students of the Tainan Community University.
6. **Steel:** The giant steel frames of the lantern now form the structure of a stormproof ball court shelter in Daxi Elementary School in Taitung. The shelter allows the basketball team to practice during rainy days.

Following its successful initial act, all parts of the project have been repurposed or recycled, thereby completing the full cradle-to-cradle material cycle as originally intended.

灯会结束后，所有构造做到 100% 回收再利用，其延伸价值包括：

1. **内环投影布幕：**由 PET 容器回收再制而成的布幕，进入第三生命周期，改制成 700 个限量环保袋，供环境教育志工使用。
2. **LED 灯：**234 组 LED 灯成为台中自然科学博物馆植物园热带雨林温室前的夜间光雕，提醒民众保护雨林生态的迫切性。
3. **竹外墙：**外环墙面 500 支桂竹转赠台东偏乡教育基地作为手工屋建材；结合太阳能光电板和雨水回收装置，在台风停水断电期间，为附近居民提供实时饮水与用电。
4. **地坪铺面：**可分解之竹筒地坪，随时间融于大自然中，就地成为公园复原的有机材料，转换成为生态养分。
5. **投影机设备工作站：**容纳投影设备的螺旋体钢结构灯座，由台南小区大学认领，结合移动式太阳能车，成为行动绿能小屋，提供民众体验绿电生活。
6. **钢结构：**主体结构 40% 为回收钢，重新组构成校园风雨球场，为长期以来只能利用教室走廊或迁就树荫作为运动空间的东部乡村小学生，提供一处不受天气影响的活动场地。

"永续之环"自装置作业到活动结束，期间所运用的构造与材料，在各阶段生命历程皆善尽其用，落实从摇篮到摇篮的可持续性价值。

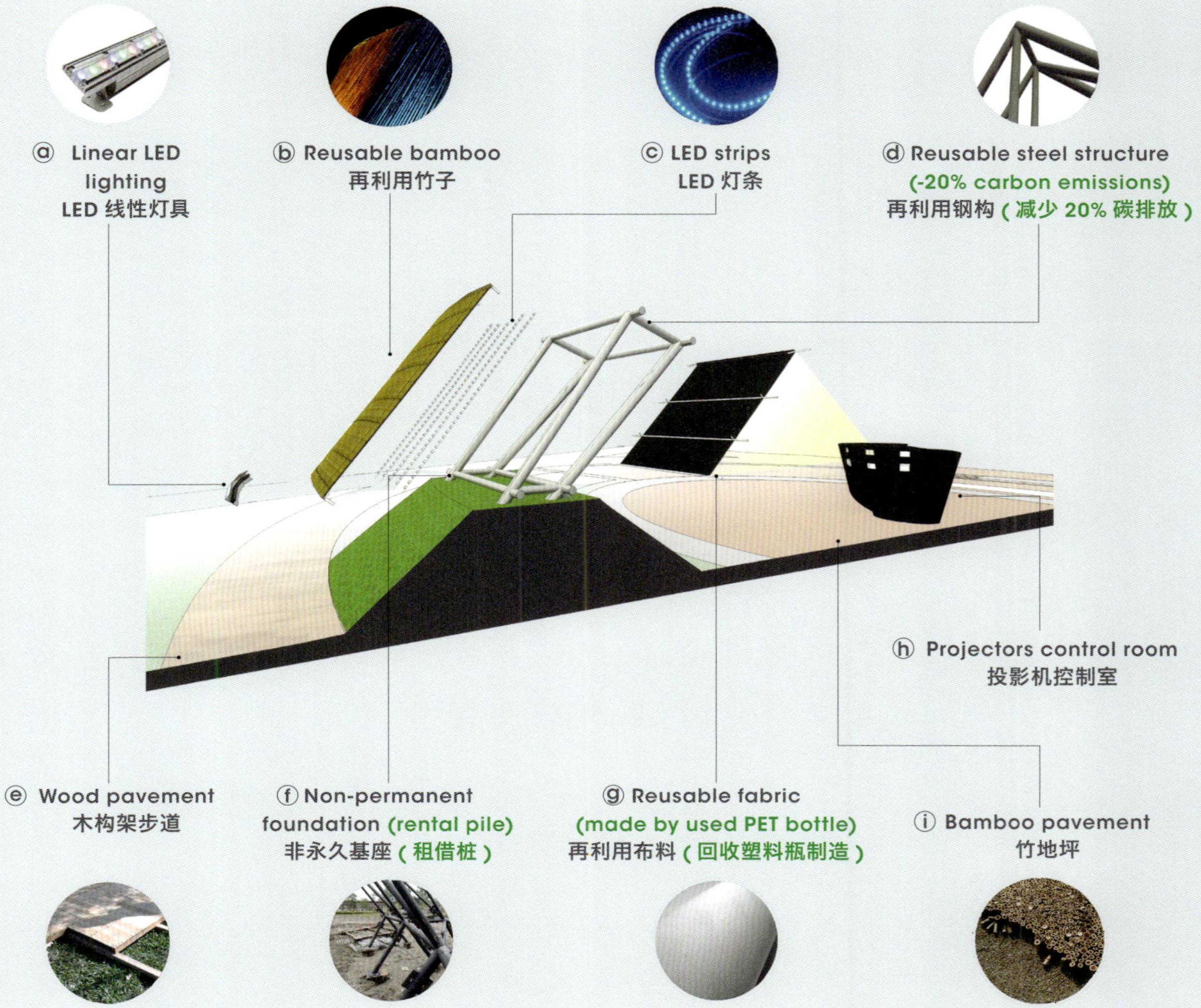

ⓖ→ Green bag 环保袋

ⓐⓒ→ Light sculpture
夜间光雕

ⓑ→ Bamboo Regeneration
桂竹再生

ⓗ→ Green energy cottage
绿能小屋

1

2

3

4

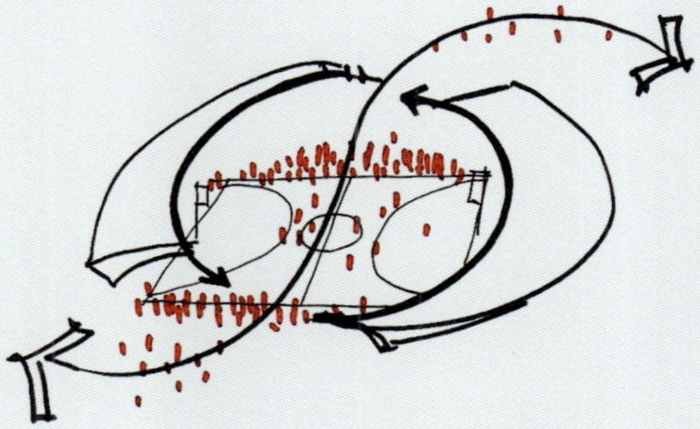

1 Construction process 施工过程
2 Bamboo pavement 竹地坪
3 Bamboo façade 竹子表皮
4 Sports court shelter of Daxi Elementary School 大溪小学风雨球场

Delta Tainan Plant Phase II
台达台南二期厂房

2010–2012
Tainan
27,487m²

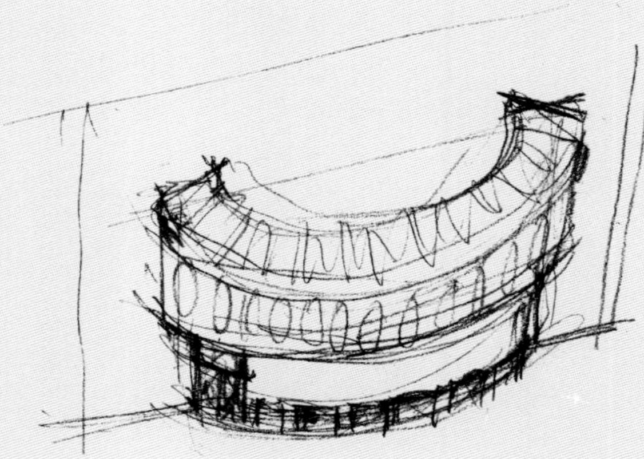

A Breathable Building
会呼吸的建筑

Throughout the project planning stages, various strategies were incorporated to allow the building to interact harmoniously with the surrounding environment, under the physical conditions of sound, light, heat, air, and water, thereby perfectly synthesizing architectural design and Delta's corporate philosophy. The design objective aims to achieve an "architecture that breathes," using buoyancy-driven ventilation and pre-cooling system strategies to lower air conditioning costs. The building follows natural solar orientation and utilizes shading devices, which lets in an adequate amount of natural light. The window openings make use of double low-e glass. The elevator system comprises of permanent magnetic synchronous elevators that preserve energy, developed by Delta. The preserved energy is recycled, and along with solar panel generated electricity, to be used for other functions within the building. All of the interior lighting utilizes LED lighting manufactured by Delta, including automatic lighting control systems. The HVAC system incorporates an ice storage central system during off-peak hours. Rain water is collected and channeled into the rainwater recycle pool, to be used for outdoor sprinkler systems and other sanitation equipment. In sum, the project not only successfully exhibits the corporate ideology of Delta Electronics, but also contributes sustainability to the Southern Taiwan Science Park through innovative green building strategies.

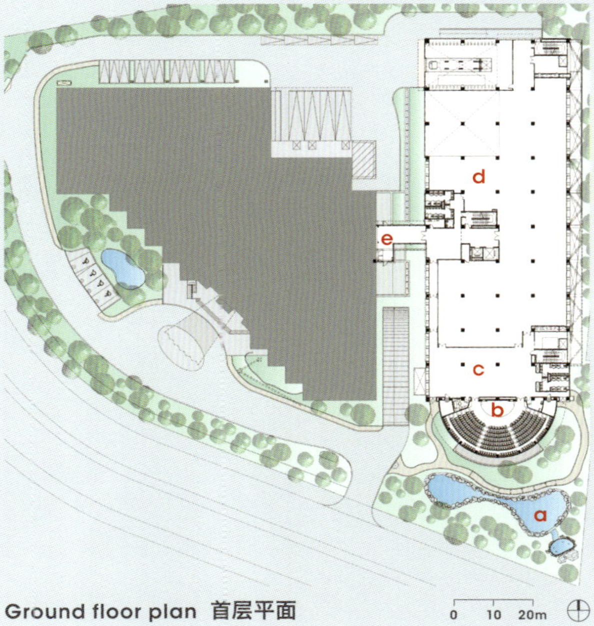

Ground floor plan 首层平面

■ Existing building 既存建筑

a Ecological pond 生态池
b Auditorium 阶梯教室
c Exhibition gallery 展厅
d R&D laboratory 研发厂房
e Connecting corridor 风雨走廊

体量规划顺应自然环境，在音、光、热、气、水的物理条件下，与周围生态和谐互动，包含研发实验厂房及阶梯教室的内部空间，均运用天然资源，满足日常运作所需，如顺应方位配置的研发实验厂房引进适度自然光及格子遮阳系统，兼顾节能与采光功效；利用浮力通风原理与空调预冷系统的 200 人阶梯教室，可享受全年自然舒适的通风环境。另外包括公共空间采用 LED 光源并加装昼光自动点灭感应装置，调整地下停车空间户外高差差距，日间室内免用人工照明；永磁同步电梯节能设计的升降设备可回收电力，与屋顶太阳能板发电产能并入供电系统；离峰储冰系统空调主机结合地道风管，进行外气预冷；以及利用屋顶平台将雨水导入筏基回收池，提供浇灌系统及部分卫生器具使用等多项节能设置及规划。一栋会呼吸的绿色建筑，如实体现企业绿能理念，也一改工业园区建筑高耗能的刻板印象。

a Shading shrubs 耐阴观叶型灌木丛
b Deck (outdoor stage) 木平台（户外舞台）
c Indoor ventilation outlet 室内出风口
d Natural ventilation inlet 浮力通风入口
e Natural ventilation outlet 室内浮力通风出口（兼自然采光）
f Landscaped seating 坐卧草坡（看台区）

Environmental control design 环控设计

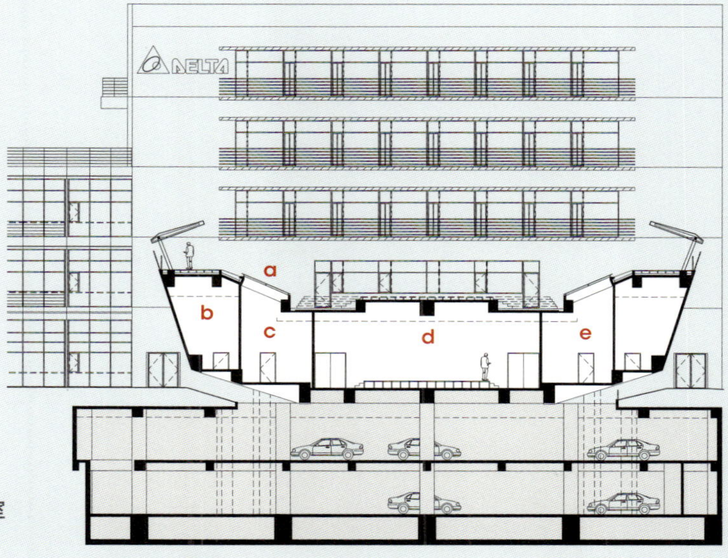

Section 剖面

a Skylight 采光天窗
b Storage 储藏室
c VIP room VIP 室
d Auditorium 阶梯教室
e Pantry room 茶水间

Delta Americas Headquarters

台达美洲总部

2011–2015
Fremont, CA, U.S.A.
16,648 m²

Net Zero Green Building
净零耗能的绿色建筑

Taking advantage of the temperate climate conditions in Northern California and incorporating a wide range of sustainable features, the Delta Americas Headquarters employs Delta's own energy-saving solutions including renewable energy, LED lighting systems, and electronic vehicle charging solutions. The building has received LEED Platinum certification and targeted to become a "net zero" green building. The project was also selected as an Honorable Mention for CBE's 2017 Livable Buildings Award for its engineering performance and exceptional energy use.

项目利用北加州地理环境优势，广泛采用对环境友善的绿色建筑设计及营造方法。透过体量方位配置、外墙遮阳，搭配再生能源、节能照明等环保节能产品的设置，将能源用量降到最低，达到建筑物全年用电量低于自行利用再生能源发电量之目标。项目不仅取得美国 LEED 绿色建筑最高等级——白金级认证，达到了净零耗能的高标准，更首创台湾团队在美取得 CBE 宜居建筑奖之国际智慧节能建筑案例。

46101

How to Net Zero
绿色建筑技术摘要

1. **Ground source heat pump**: This system uses a loop field of pipes underground that transfers heat to or from the ground to boost efficiency and reduce the operational costs of the heating and cooling systems.
2. **Radiant floors:** A network of pipes embedded in the concrete floor circulate water to provide radiant heating or cooling.
3. **Solar energy system:** The building roof surfaces and a large canopy above the lobby volume supports an extensive PV solar system with inverters of 98% efficiency.
4. **Natural ventilation:** The exterior wall features operable windows and a central atrium along with two corner towers in the offices act as thermal ventilation stacks.
5. **Active chilled beam:** The active chilled beams in the enclosed conference rooms or offices provide for individual room controls in an energy efficient manner.
6. **Sun shading:** The building is aligned with true north and the south elevation features horizontal brise-soleil, while the east and west elevations have deep recessed windows.
7. **High performance building envelope:** The building exterior wall features R-30 insulation in the walls, R-45 insulation in the roof, and low-e double glazing.
8. **Landscaped open space:** The project preserves at least 50% of the site as open space with native landscaping plants that will help support wild life.
9. **Ecological retention pond:** A bio-retention pond with native low-water consumption planting is incorporated.
10. **Rainwater harvesting:** A 140,000-gallon rainwater collection system is installed to provide water source for landscape irrigation.

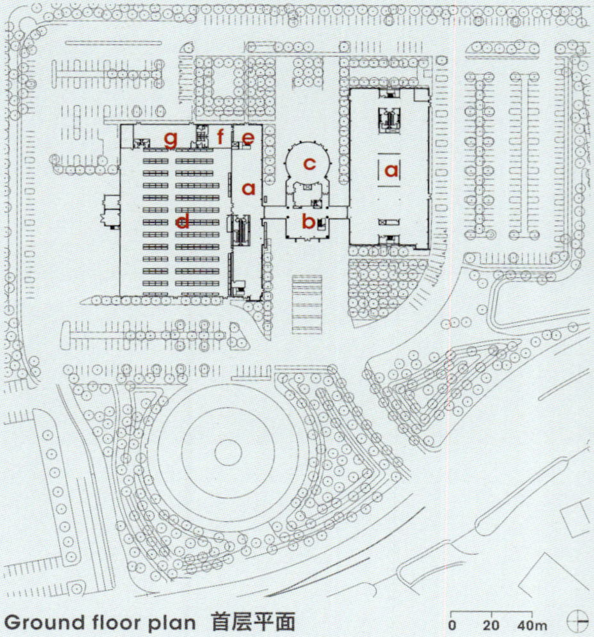

a Open office 开放办公室
b Lobby 大厅
c Multi-purpose room 多功能空间
d Warehouse 仓库
e Training 训练教室
f Mechanical 机械室
g Warehouse office 仓库办公室

Ground floor plan 首层平面

1. **地下水源热泵**：北加州冬季地下水温比室外气温高，夏季地下水温低于室外，利用土壤中铺设的水管进行热交换。布满室内地板的空调水管，对上下楼层均可产生空调效果。
2. **冷热辐射地板**：利用混凝土极佳的热惰特质制成，将管道预埋在楼板间。夏、冬两季分别以不同温度的冷热循环水降温、供暖，成为控制室温的主要媒介。
3. **太阳能系统**：采用高效能太阳能面板及高达 98% 效率的太阳能转换器。
4. **自然通风**：利用当地良好气候条件广设开窗面。部分建筑端角拉高作为浮力通风塔。
5. **主动式冷梁**：运用于办公区较小隔间的空调系统，是一种集制冷、供热和通风功能为一体的空调系统，提供良好的室内气候环境及单独区域的温控机制。
6. **遮阳**：大楼中庭开窗，南面采用水平遮阳板，东西向则以深开窗方式，让光线穿透，室内均匀采光。
7. **高性能外墙**：采用光谱选择玻璃，将热反射于建筑外，降低耗能。
8. **绿地**：基地保留百分之五十以上的面积作为绿地，并栽种原生植栽，丰富生态多样性。
9. **生态池**：提供生态自然调节功能，改变建筑物附近的微气候。
10. **雨水回收**：装设 14 万加仑雨水回收系统，用于浇灌植物，应对当地终年少雨、干旱及缺水危机。

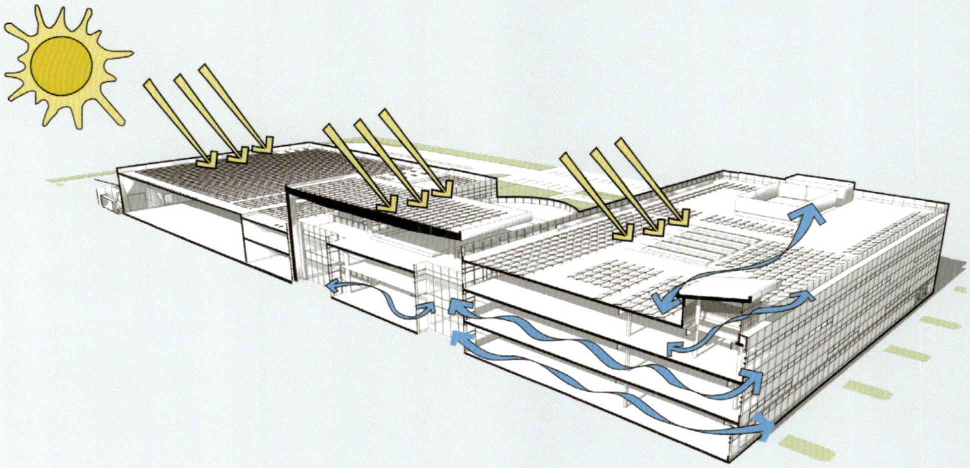

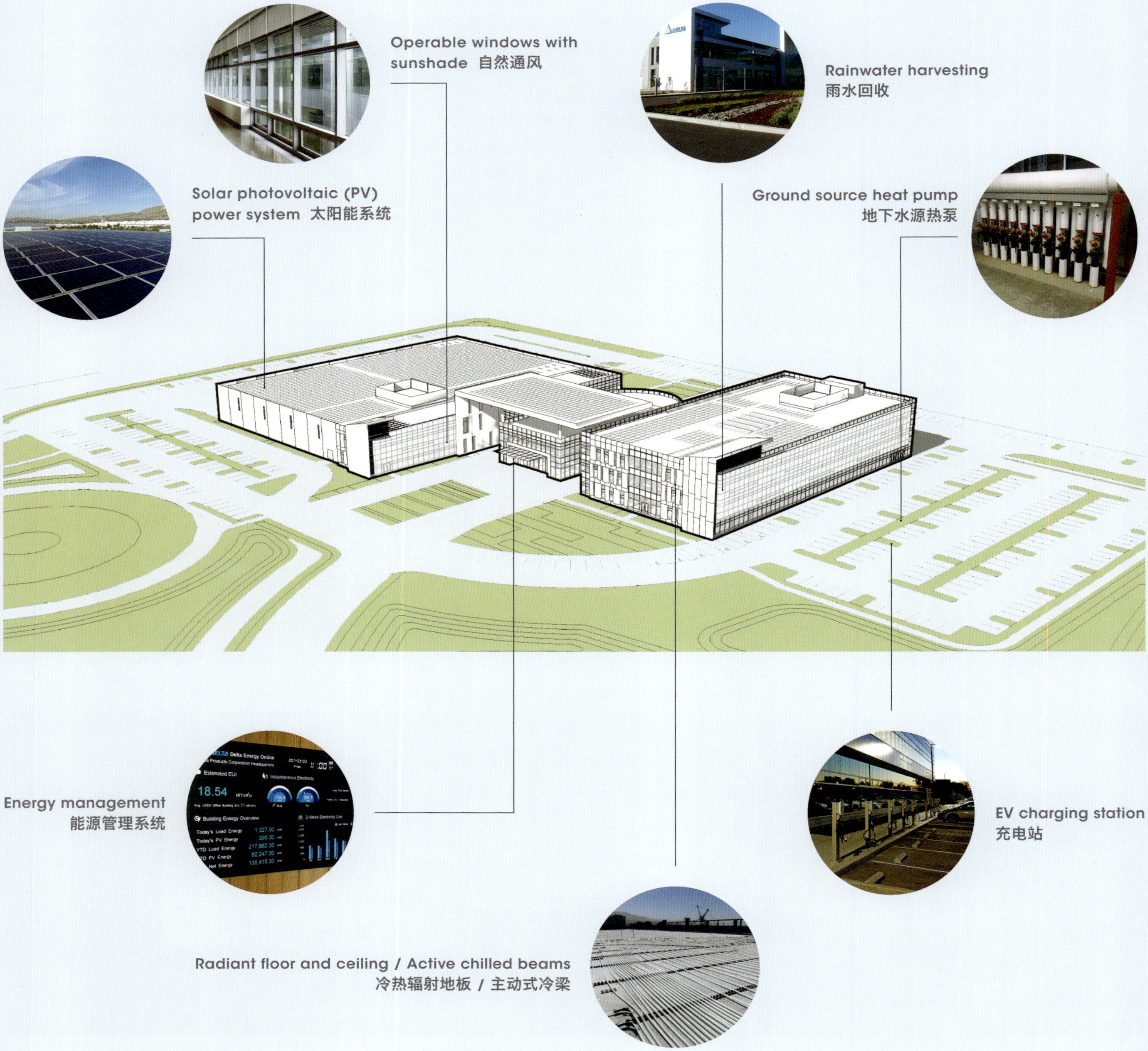

Operable windows with sunshade 自然通风
Rainwater harvesting 雨水回收
Solar photovoltaic (PV) power system 太阳能系统
Ground source heat pump 地下水源热泵
Energy management 能源管理系统
EV charging station 充电站
Radiant floor and ceiling / Active chilled beams 冷热辐射地板 / 主动式冷梁

1

2

3

1 Office 办公室
2 Office atrium 办公区中庭
3 Warehouse 仓库

Taichung First High School Heritage Building
台中一中校史馆修复及再利用

2012–2015
Taichung
822 m²

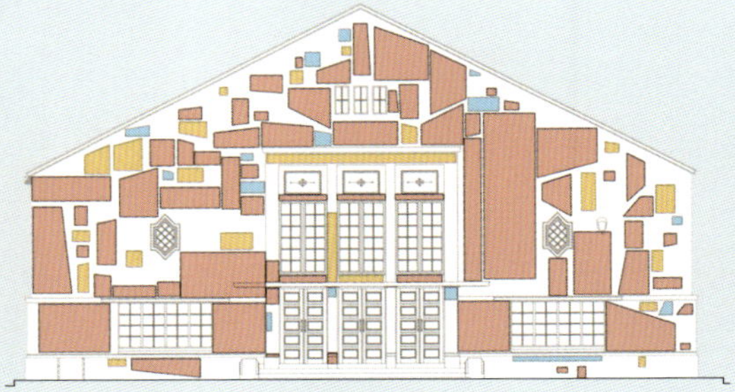

Damage survey
损坏程度

- Critical 严重
- Significant 中度
- Minor 轻度
- Minimal 最小

Reactivate a Unique Setting
活化历史场所

Completed in 1937 during the Japanese occupation of Taiwan, the historical School Heritage Building of Taichung First High School is the only remaining building from the founding era on campus. The renovation was expected to utilize up-to-date sustainable design features to serve as a venue for green building advocacy and environmental education. The refurbishment, completed in 2016, has successfully met the design challenge.

Functionally, the Heritage Building was originally intended for Japanese Shinto spiritual education and subsequently used as an auditorium and student activity center. In the restoration, the Shinto Shrine chamber at the north end is repurposed for extensive performance functions while the exhibition contents are relocated to a pair of new balconies at the south end of the building that can be reached via stairs or a hydraulic lift. The connecting walkway between the two balconies creates a dramatic opening above the main entrance doors. After decades of neglect, most of the original architectural details and authentic finishes were either removed or concealed. To restore the glory of the Heritage Building as a spiritual center of the school, all subsequently added partitions and ceilings were removed. The original steel "Fink truss" was then refurbished so that the delicate roof structure became a significant architecture feature. On the exterior walls, multiple layers of improperly applied paint coatings have been carefully removed, and then a certified finish compound in the building's original color is applied.

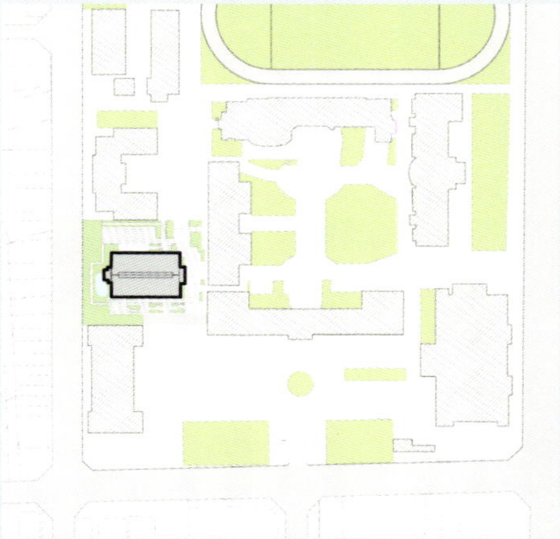

Campus plan 校园总平面

Before

第一中学校讲堂是日据时期台湾人创建学府中的仅存建筑物，具有历史传承及文化教育上的实质价值。随时空变迁、校务设施扩建，讲堂演变为史料展示空间及闲置仓库，因此，项目修复的三项课题定为保存文化资产实体、维护历史意涵价值、塑造叙事场所。希望透过重新定义与定位的校史馆，不仅成为社会教育及文化传承的媒介，更是一个见证历史的最佳场所。

为活络历史空间机能，增建二楼钢结构平台作为校史文物展示区及活动观赏看台之用，改变礼堂内观众的观看角度，以及与表演者之间的互动模式，注入新的使用形态。配置维持原建筑物的形制和中轴特性，二楼增建构造延续原屋顶结构“芬克式桁架”的钢结构美感，新旧构造间刻意保留 5 厘米隙缝，呈现出新旧分明的对照，并开启上下层空间的对话。此外，修复团队全面清查墙面损坏状况，依实际状况进行不同程度修复，并局部保留整修过程中经由剥漆而露出的不同年代面饰材，成为就地观察历史痕迹的最佳教材，重塑历史场域新风貌。

Historical Building Turned Green
史迹添绿意

In an effort to promote sustainability, the linear sky-lights incorporated air vents at the roof ridge, which filled the interiors with diffused natural light while discharging warm air. In addition, the open space around the building is re-landscaped with shade trees and shrubbery to harmonize with the existing school yard. Specifically, the western and southern sides are densely planted so that cool air accumulated underneath the trees are pulled in through the restored lower sash windows while warm air is being drawn out through the upper ventilation windows and ridge vents. Furthermore, electricity generated by photovoltaic panels on the roof of an adjacent building supplies all of the building needs. The incorporation of renewable energy together with improved natural lighting and air circulation has helped the project achieve its neutral energy goal.

为推广可持续的负排放理念，重新设计屋顶通风构造，提升室内浮力通风作用；利用屋脊下方梭形的灯具蓄热区产生温差，增加空气对流，同时提供间接光源，降低室内辐射热能；另设置太阳光电板，积极落实节能永续。重新规划后，校史馆周边绿地整合开放空间、景观地坪及水池，成为校区内文化遗产教育场域；将校园围墙改建为透空形式，对外吸引市民贴近史迹文物，提升建筑文化资产的再利用价值。

Before

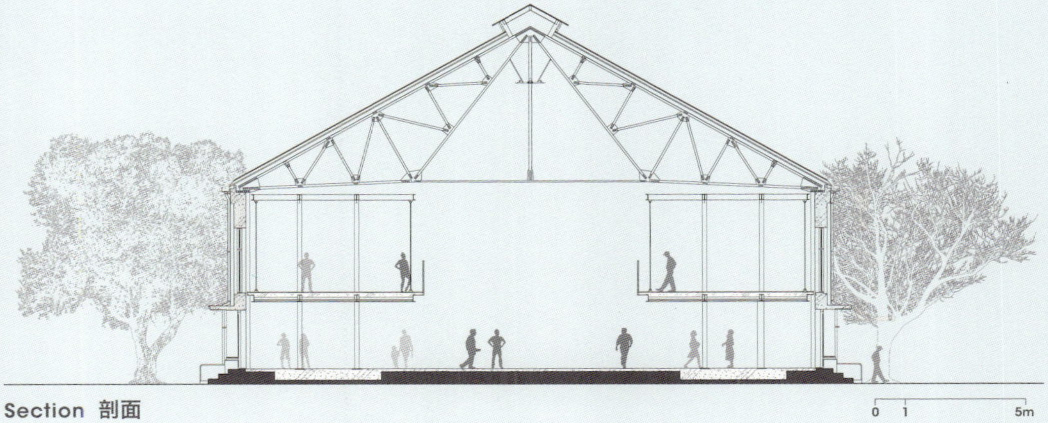

Section 剖面

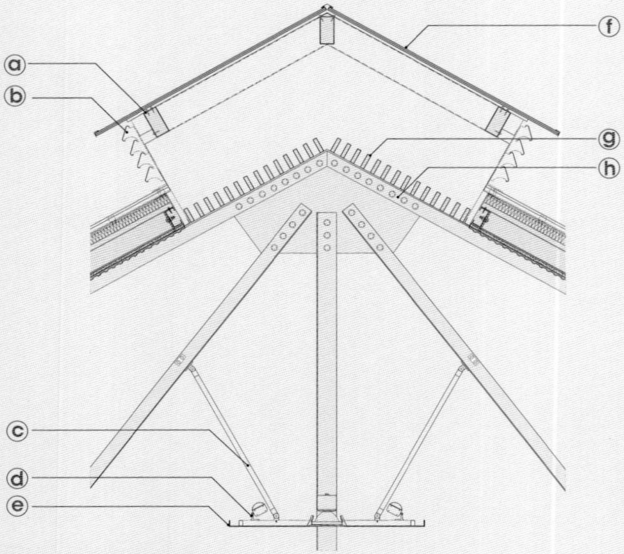

Detail of roof ridge
天窗及通风气屋脊细部

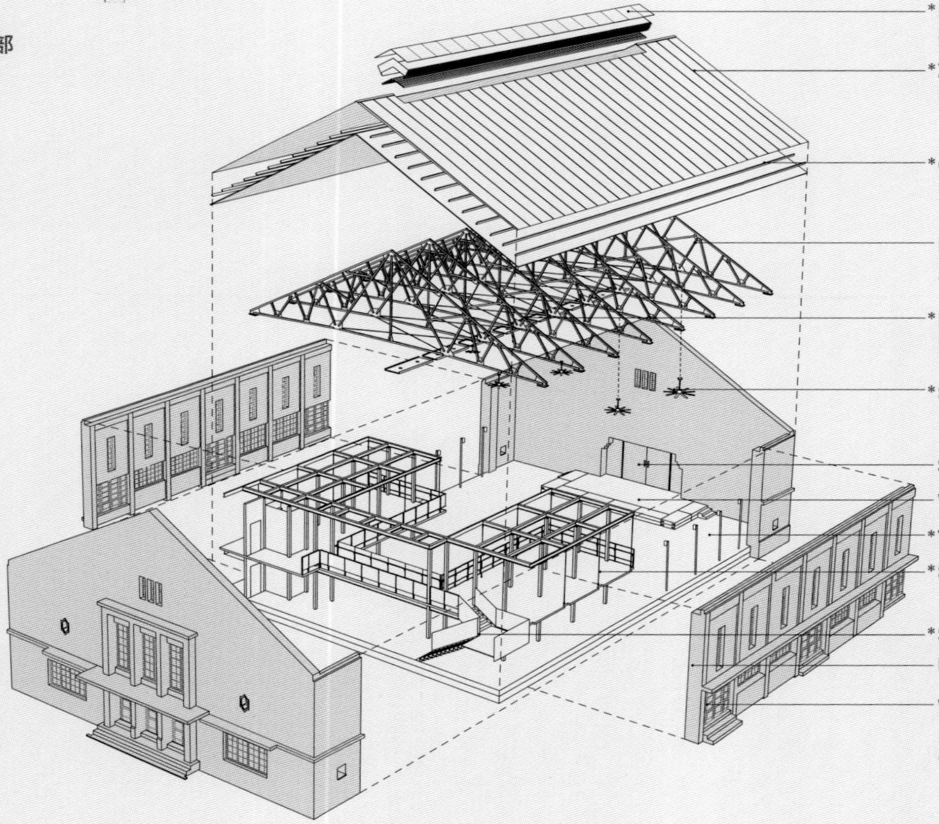

Architectural assembly diagram
建筑构造分解图

Horizon Design Shanghai Office
群裕设计咨询上海办公室

2005
Shanghai
700 m²

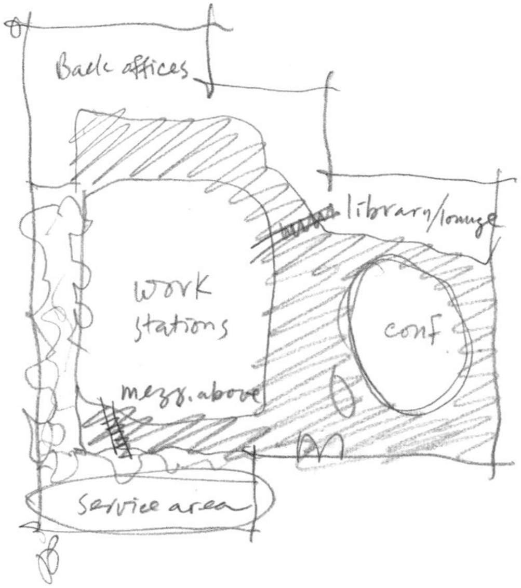

Build the New with the Old
旧材新构

Situated in the Yangpu Creative Industrial Park, the site of Horizon Design's Shanghai Office served as one of General Electric's largest manufacturing facilities in Asia in the early 20th century. Having witnessed Shanghai's industrial development, the building is of significant historical value. The renovations create a symbiosis between the historic and modern elements of the building, breathing new life into the structure by preserving traces of the different ages while using modern building materials. For example, the steel frame of the original entrance canopy is covered by semi-transparent polycarbonate prefabricated panels and dry reeds.

设于上海市杨浦区滨江创意产业园区一隅的群裕办公室，初为 1921 年美国通用电气在亚洲设立的最具规模的电子工厂，是极具历史价值的工业文化遗址。设计策略上除保留建筑原貌外，取用基地废弃旧建材，并添加崭新的材料语汇，在原入口雨篷钢架上方，采用半透明聚碳酸酯预制板和干芦苇加盖顶篷。钢和玻璃的光滑与原建筑表面的粗犷感形成对比，以传递历史与现代接轨的时代意涵。

A Dialogue Between the Old and the New
重组新旧空间的关系

The workspace consists of a series of "mini-environments" of various shapes. These "buildings within a building" help shape another order of spaces under a single roof, interacting with each other to initiate a dialogue between the "old" and "new" elements through an analogy of streets, plazas and other building blocks in a city. The full-height glass partitions wrapping the studio spaces help alleviate pressure from the roof's massive volume. A lightweight steel staircase neatly connects the studio above to a triangular stage constructed with recovered wood planks serving as a library. The conference room forms an elliptical shape with a solid wall made of locally reclaimed bricks. For major events, the conference room's translucent partitions can be rearranged to provide a visual and physical connection between the library and conference room. In sum, Horizon Design serves as an exemplary precedent establishing a dialogue between old and new, between preservation and renewal, and between the traditional and the fashionable.

经由旧建筑再利用，企图在现代感与国际化的潮流中，以“建筑中的建筑”之形态交错配置其中，顺势围塑出有如街道与广场的室内空间。大玻璃盒般围合的工作区，上方夹层为展示空间，透过轻盈的钢结构空中天桥衔接图书平台，利用回收望砖砌筑椭圆形会议室，当活动隔扇开启时，会议室与图书区分界消失，成为可弹性使用的舞台和观演席，让使用者与新、旧空间互动交融，在此微型城市般的记忆平台里，透过工作、讨论与生活，让历史与当下有机接续，传承对话。

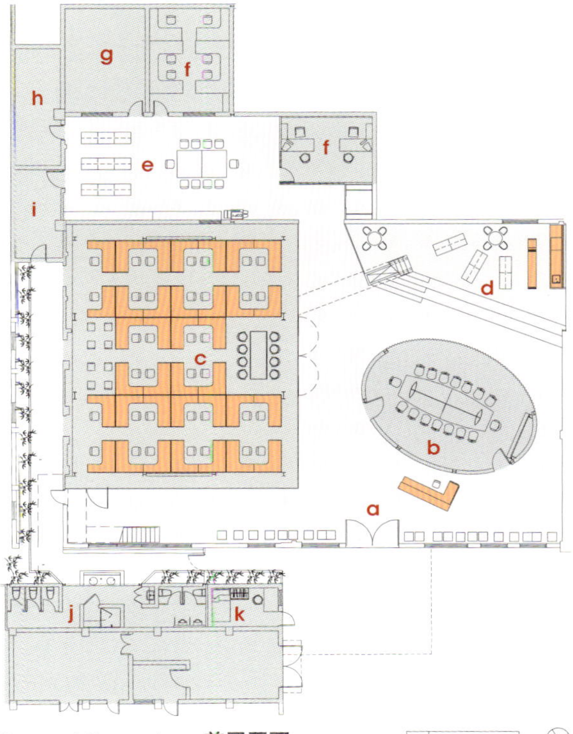

Ground floor plan 首层平面

a Lobby/Exhibition 大厅／展览室
b Conference room 会议室
c Studio 工作室
d Lounge/Library 会客室／图书馆
e Sample/Model room 模型室
f Administration 行政办公室
g Document room 资料室
h Storage 仓储
i Mechanical 机械室
j Restroom/Shower room 卫浴室
k Guard room 管理室

Building block 建筑街区
Individual space 个人空间

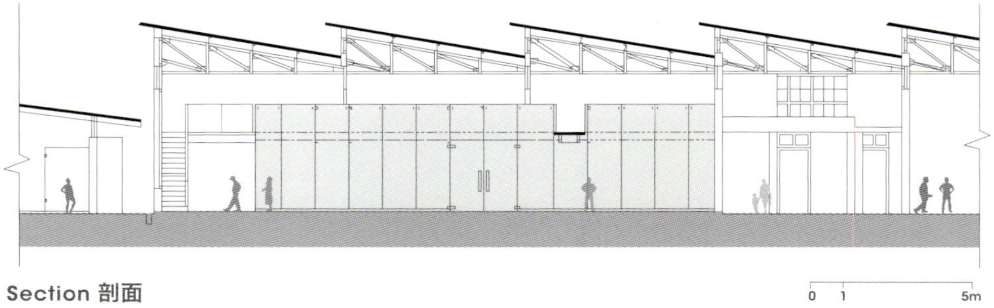

Section 剖面

Horizon

tsmc Campus
台积电园区

Hsinchu Campus/ 1993–
Southern Taiwan Campus/ 1997–
Central Taiwan Campus/ 2009–

ELP:
Ecology, Life, Production
三生计划:
生态、生活、生产

As a leader in semiconductor technology, the Taiwan Semiconductor Manufacturing Company (tsmc) has been working together with JJP to create a sustainable and ecological benchmark for the next generation of industrial architecture. The ELP plan—ecology, life, production—came forth as a design guideline for campuses in the Hsinchu, the Central Taiwan Science Park, and the Southern Taiwan Science Park.

居于全球半导体技术领先地位的台积电公司，长期以来与 JJP 设计团队共同致力实践可持续的建筑理念，于台湾北中南三地科学园区打造生态、生活、生产并重，全方位绿化节能的新世代生态厂区。

Hsinchu Campus-
Fab 12, Phase IV &
Headquarters
竹科十二厂区总部大楼、
四期

Carried out in seven phases, the Hsinchu campus was Taiwan's first 12-inch wafer foundry. A progressive and energetic formal visual language was created to reflect the firm's position as the driving engine of the industry. The building complex is arranged from north to south according to functional requirements, while addressing the trapezoidal shaped site. The main lobby of the Phase I buildings is located at an off angle with the Headquarters Building, recalling the irregular boundary at the south of the site, while a variety of terraces, water fountains, walkways and light

台灣積體電路
Taiwan Semiconductor
Manufacturing Company, Ltd

wells are incorporated to integrate indoor and outdoor spaces and provide a pleasant visual extension. The design concept for the Phase IV building was to create a human-centered "green city". The setbacks along neighboring sites are purposed as ecological retention ponds. The two main buildings are linked by an enclosed bridge to transport materials within a controlled environment, which enhance the effective movement of people and goods while reducing the environmental impact.

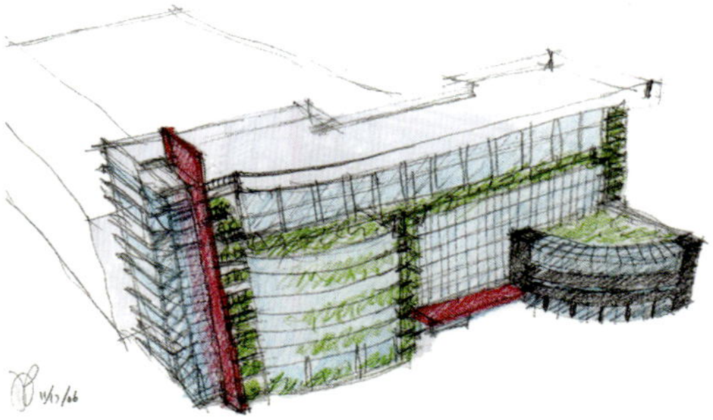

台积电十二厂为台湾首建 12 英寸晶圆厂，以具火车头意象的总部大楼为首，历经七期扩建完成厂区建置。梯形基地由北向南排列，一期办公栋东北侧规划成为主要景观区及总部主入口广场，衬托总部意象。四周借由平台、水景、步道、天井等元素，表达对周边环境的尊重；四期以人性化绿能概念营造一座"新绿之城"，取代传统效率至上的制造思维，如护城河般的生态池，运用与邻地衔接的退缩绿化带，设置生态水道，丰富基地生物多样性；位于邻近基地的两主厂以无尘、洁净自动输送系统的"连通管桥"作跨区连接，降低环境负荷。十二厂区不仅传递出永续经营理念，也为绿色生态厂房奠定了示范基准。

Atrium section 中庭剖面

Southern Taiwan Campus- Fab 14, Phase V
南科十四厂区五期

The idea of the ELP plan—ecology, life, production—was initiated in the early planning stages of the Southern Taiwan campus. Since the project is surrounded by three retention ponds, the façade of the office building features a theme of "sparkling ripples," to echo the beautiful lake scenery. Vertical insulated metal panels with different shades of grey are employed on the planting walls to create a rhythm on the immense walls, and help soften the scale of the building. The project also employs forward-thinking sustainable design strategies such as retention ponds, multiple layers of native vegetation, green roofs, grey water recycling, and rain water harvesting. The project has achieved EEWH Diamond and LEED Gold certification.

十四厂在初期规划阶段，即传递生态、生活、生产的主题信息，五期兴建则完整落实了“三生”平衡的可持续性概念。本期建筑将基地特殊水域景观纳入设计元素，办公栋以遮阳立面诠释周边三座滞洪池面波光粼粼的景致；厂房则利用外墙垂直复合板错落变化的深浅色阶创造律动感，随视觉角度转换呈现水波闪烁般效果，降低庞大体量之冲击；另外运用景观生态池与生态河道、复层式植栽、生态绿岛、屋顶绿化及雨水回收再利用等设计手法，创造多孔隙的生态环境，提供多样化的生物栖息空间。在率先取得的台湾钻石级绿色建筑及美国绿色建筑 LEED 黄金级双认证后，十四厂设计规划更趋完善，成为绿色生态厂房标杆。

1

1 Façade and ecological pond 表皮与生态池
2 Sparkling night view 波光粼粼的夜景

2

Central Taiwan Campus-Fab 15, Phase I
中科十五厂区一期

Located on a hillside in the Central Taiwan Science Park, the project integrates a 40-meter wide greenbelt, a central atrium, and screening systems to enable an interaction between users and nature. The office building incorporates a vertical unit with horizontal bars split by the central atrium, which consolidates circulation and provides natural ventilation through the top. The façade features screening panels for shading and greenery. An eco-pond and strolling paths wrap around the site, and have seen the return of fireflies and butterflies, which are signs of environmental recovery. With the elements of nature emanating from the screening panels to the atrium and the greenbelt, the project goes beyond simply meeting requirements to establish a new standard of green factory meant for work, people and the environment.

位于山边的十五厂区视野极佳，可远眺台中盆地；建筑立面与内部空间以绿意串联，40 米退缩绿化带将自然元素延引入工作环境；办公区绿化中庭与高楼层配置休闲平台，提供员工交流及休憩空间。垂直绿化加设太阳能板遮阳，兼具绿化节能；生态水池与生物廊道成功重新孵育萤火虫、紫斑蝶等生态指标动植物；绿意满载的厂区，在产值达标之余，更积极推广生态绿色教育，建构修补、复育、复原的永续科技人文园区。

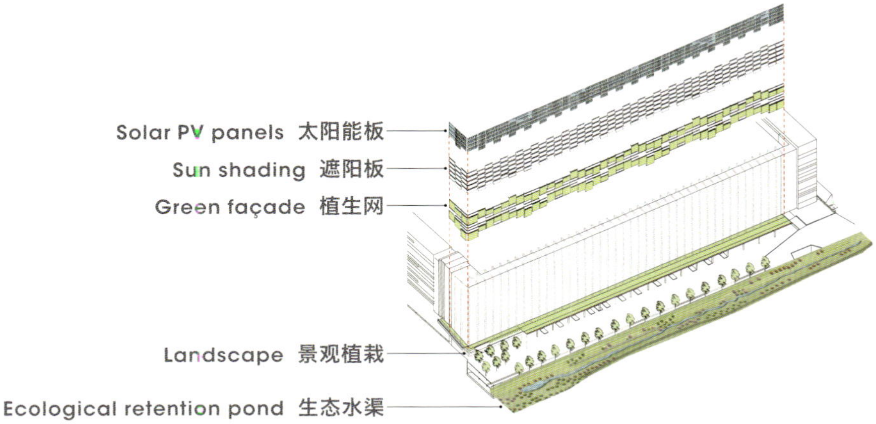
Solar PV panels 太阳能板
Sun shading 遮阳板
Green façade 植生网
Landscape 景观植栽
Ecological retention pond 生态水渠

Aging Gracefully: Suang-Lien Series and Interdisciplinary Workshop
从熟龄到乐龄：双连赡养系列与熟龄好宅研究计划

1

Suang-Lien Senior Living Community
双连教会社会福利园区

1997–2009	
Phase I	8,670m²
Phase II	13,100m²
Phase III	1,522m²
Church	995m²

2

Suang-Lien Xinzhuang Social Welfare Center
双连新庄社会福利中心

2013–2015	2,040m²

3

Interdisciplinary Workshop: Next Generation of Senior Living
新世代熟龄好宅研究计划

2016–2017 Research Project

Aging population has been a global issue for years. In over two decades of designing for the elderly, the JJP team has continued to advance the field through data collection and the application of empirical evidence on the relationship between environment and behavior, in order to provide a warm and meaningful place for the elderly and to enhance their quality of life. Projects undertaken for the Suang-Lien Presbyterian Church have been particularly representative. Recently, in response to changes in social structure, the JJP team is exploring the evolutionary possibilities of architectural prototypes that go beyond the original residential building types and connect with sustainable living to the future. The result is the "Next Generation of Senior Living" interdisciplinary study.

人口老龄化已是世界各国共同面临的社会课题之一。JJP 团队逾 20 年关注于老人赡养、长期照护的机构式建筑实践，透过使用者与实质环境互动所产生的场所经验，持续以反馈、实证、精进设计，为长者规划温暖且具有意义的场所，提升银发住居质量，其中尤以与双连赡养中心合作案例为代表；近期更积极以建筑思维响应社会结构转变，探讨原居安老的建筑应用原型与永续生活的可能样貌，推生跨界研究“新世代熟龄好宅＂计划。

Suang-Lien Senior Living Community
双连教会社会福利园区

1997-2009
New Taipei City

Rebuild the Sense of Belonging
重塑认同感

Amidst spectacular scenery, the construction of the community was carried out in phases, including residential living space, memory care and church facilities. The consideration of factors such as local climate and the circulation of different care types in this continuing care retirement community (CCRC), led to the adoption of a linear extension in order to achieve a continuous green space. Based on the study of the environment-behavior relationship, a successful, supportive senior living environment was created by designing spaces that provide psychological, physical, and social support to the elderly residents. The environment creates the sense of belonging, and engenders a sense of home.

基地背山面海，风景秀丽。园区整体规划分期分区开发，包括赡养、养护、失智照护及教堂等设施。考虑当地气候与多层级连续性的照顾形态及服务设施的衔接等因素，园区体量以线性延伸配置，建构出完整舒适的绿地空间。JJP 团队透过建筑空间与行为之间关系的研究，在设置无障碍多功能环境硬件设施、提升舒适生活形态的同时，亦考虑到长者对晚年迁移与旧居情感依附可能产生的不适应感，因此借由空间设计建立“家”的认知，提供居住者一个心理、身体及社交生活全方位支持的环境，重塑认同感，酝酿回家氛围。

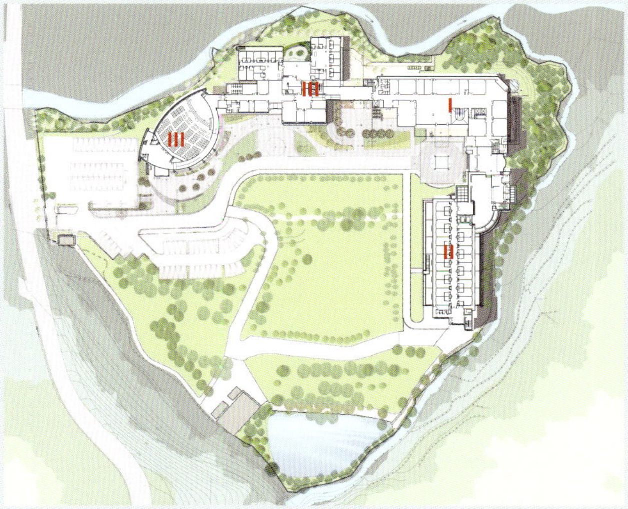

Site plan 总平面

Phase I: 8,670 m²
Phase II: 13,100 m²
Phase III: 1,522 m²
Church: 995 m²

2

3

1

The Phase I building serves seniors who are healthy as well as those who require further care and assistance. In order to create a sense of home, spaces like corridors are conceived as interior streets; while residents' entry recesses, or bay-type spaces are conceived as neighborhood corridors or front porches. Together these elements set the stage for informal contact and a sense of inclusion to support residents' social needs.

Phase I and II buildings are connected by public spaces such as senior classrooms and cafes. Furthermore, a lounge is provided on every floor while the main dining room is situated on the seventh floor for the best view.

Phase III of the community is the first elderly care facility in Taiwan that adopts the Unit Care system for the care of people with dementia. The central service space, nursing station, observation rooms and physical therapy rooms occupy the core of the building. Each care unit has direct access to the therapeutic garden. Humane care methods and facilities like these have made the Suang-Lien Senior Living Community the first choice for visits from cross-strait senior care agencies.

首期的赡养及长春照护中心，主要服务生活可自理及需看护的长者。为营造家的归属感，在各房门前设计“前院”及“街道”的空间意象，唤起长者们过去“串门子”的生活记忆。

二期延续一期规划，透过教室、咖啡厅、健康促进区等公共空间串联衔接一期，提供长者终身学习及活动交谊空间，并于七楼最佳海景处设置餐厅，提供舒适的用膳环境。

第三期为台湾首创的单元照护模式的失智症照护专区，核心为中央服务空间、护理站与共同活动区，各照护单元透过定制化机能设施，可直接通往疗愈花园。园区逐期完善的相关设施以及强调人性化的照护方法，已成为两岸赡养机构参访的首选目标。

4

5

1 Public space 公共空间
2 Lounge 大厅
3 Elevator 电梯
4 Public corridor 公共廊道
5 Living room 客厅

A Calming Sanctuary
心灵安歇的帐幕

The symbol of the "fish" in the Christian faith and the imagery of the "tent" inspired the design concept of the church building. The building is shaped by the site's prevailing wind direction and land boundary to form two curves in the form of a fish figure. The steel frame rises up along the two curved planes to complete an arched structure. The church faces the development's vast expanse of lawn. The front rows at the ground level accommodate those using wheelchairs, while the side passageway connects to the lobby shared with the Dementia Care Zone. The furniture was tailor-made with the physical constraints of the elderly in mind. At night, the light cast on the curved church walls projects the image of a tent between heaven and earth, with the spiritual symbol of the cross on top. It is hoped the church will become a warm and soothing sanctuary that gently comforts the hearts of visitors.

礼拜堂构想配合曲折的河沟及地界，延引“扩张帐幕、张大幔子”概念，在室内顶部交织出基督教常见的鱼形符号。入口朝南面向园区草坪，地面层观众席前排规划轮椅专区，侧边开敞走道衔接大厅与厕所及茶水间，顺应长者体力及需求，方便进出。家具尺寸、材质皆透过实际模拟测试，符合行动不便长者的使用需求，另考虑易清理，利于长者起身、挂拐杖等细节来定制家具。夜间运用灯光投射大片弧形墙面，并以单点投光扫向十字框架，凸显精神目标，让园区内长者与当地住民拥有一个身心灵安歇之处。

Suang-Lien Xinzhuang Social Welfare Center

双连新庄社会福利中心

2013–2015
New Taipei City
2,040m²

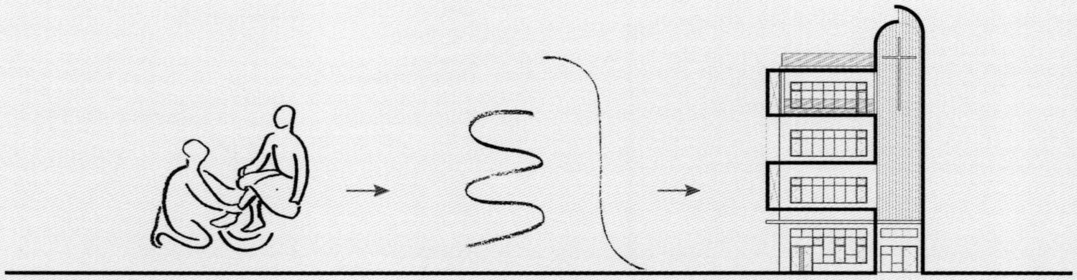

Social Welfare in Communities
小区照护

Located in a fast-growing residential area, the Xinzhuang Center is an exemplar of community daycare for the elderly that provides health rehabilitation, daycare services, meal delivery, house visits and consultation by phone when required. The first floor is a clinic for health rehabilitation, the second and third floors provide daycare services for the elderly along with a training center, and the fourth floor has a public dining room and kitchen. The open-air terraces on the second and third floors feature therapeutic gardens, where people may sit and chat, or take a walk among lush greenery, or enjoy gardening using planters with open bottoms designed for wheelchair access.

项目位于快速发展的住宅区，提供小区及居家老人餐饮外送、亲临访视、电话问安、健康促进及居家照顾等服务，属于区域性照顾服务中心，为该地区老人日间照护医疗发展的重要据点。一楼规划为康复科诊所，挑高的地面层提供明亮流畅的候诊与复健空间；二、三楼分别为日间照顾及人才培育中心，四楼设置中央厨房及餐厅。体量利用雕塑手法形式退缩露台空间，设置疗愈空中花园，塔楼格栅的夜间透光效果，在栉比鳞次的都会丛林区中点亮一处温馨。

1

2

1 Clinic 诊所
2 Healing garden 疗愈花园

Interdisciplinary Workshop: Next Generation of Senior Living

新世代熟龄好宅研究计划

2016–2017

An Era of Value Co-creation
价值共创的新时代

Not simply content with over 20 years of experience in healthcare design, JJP invited numerous experts in the real estate and healthcare industries to conduct an interdisciplinary workshop on the future of healthcare design: The Next Generation of Senior Living. The methodology of this study was to cultivate the prototype of a "Behavioral Building" through four stages of study: interviewing, researching, behavior & psychological observation, plus 2D and 3D simulation to bring about innovations in the spatial design. The prototype of the "Behavioral Building" would describe how the users might change and/or interact with the building's scale, units, interior design based on their ages, ranging from 60 to 90. The study indicated that the traditional residential mode cannot meet contemporary users' needs due to evolving social conditions such as changes in family size or social values. Hence, the study engaged in "User Research" to aid the development of "Rapid Prototyping," as well as the search for the potential values of the "Next Generation of Senior Living".

JJP 团队从多年医养照护设计经验，进一步以创新思维出发，跨界邀请建筑实践团队以及通用设计等执行伙伴，共同进行熟龄好宅研究计划。透过各种类型的访谈、研究、行为与心理的观察、社会议题的导入、二维与三维的模拟以及空间设计上的创新，建构出“行为建筑”的演化结构；借由多个由使用者行为与建筑构件彼此影响互动的“行为式建筑架构”，依据熟龄阶段演变规律调整建筑空间与用户活动。研究显示，因家庭组成与社群观等社会条件转变，传统住宅模式已不敷时代需求，故研究计划通过使用者研究以助于“快速原型生产”，企图挖掘出未来熟龄好宅之潜在价值。

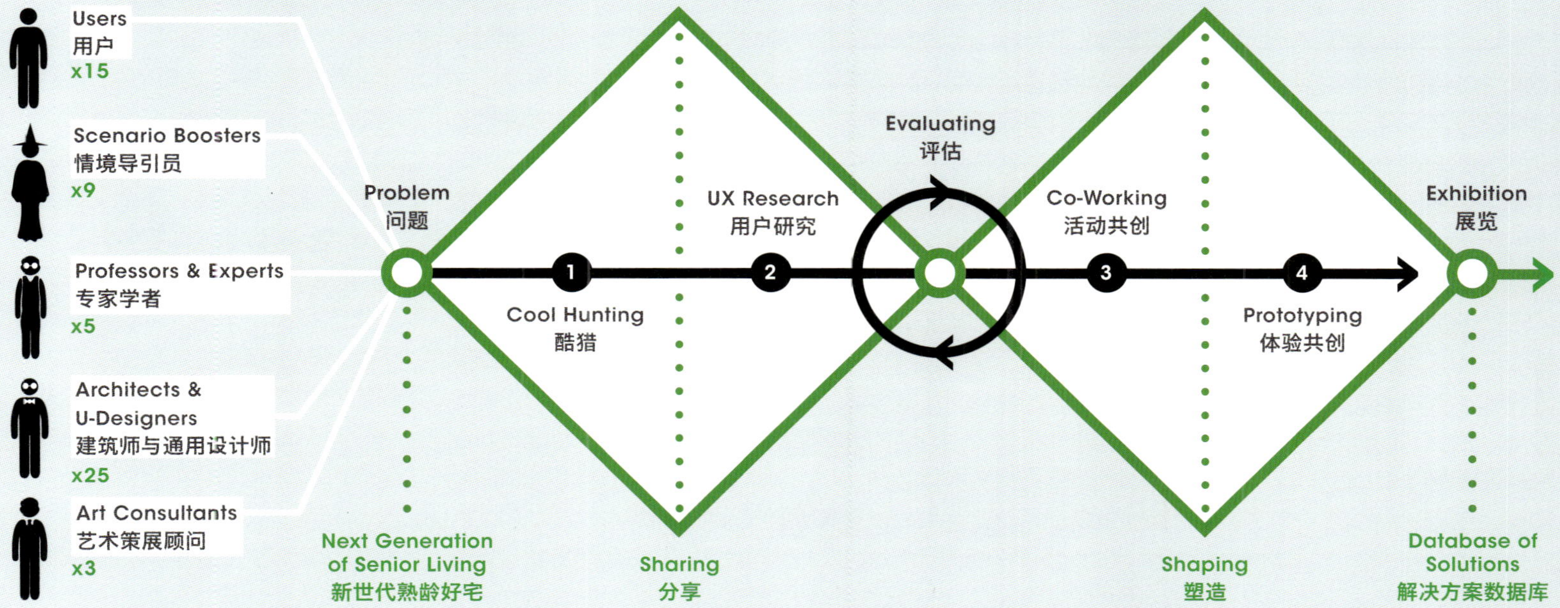

Users
用户
x15
Scenario Boosters
情境导引员
x9
Professors & Experts
专家学者
x5
Architects &
U-Designers
建筑师与通用设计师
x25
Art Consultants
艺术策展顾问
x3
Problem
问题
Cool Hunting
酷猎
UX Research
用户研究
Evaluating
评估
Co-Working
活动共创
Prototyping
体验共创
Exhibition
展览
Next Generation
of Senior Living
新世代熟龄好宅
Sharing
分享
Shaping
塑造
Database of
Solutions
解决方案数据库

Scenario-based Methodology
情境导引行为式设计研究方法

The study was divided into four stages/concepts: Scenario, Users, Activity, and Experiences, and then gradually grew into different aspects of the state-of-the-art values of the Next Generation of Senior Living.

研究计划共分为四个步骤：情境研究、用户研究、活动共创、体验共创，以多面向的使用者研究方法，逐步扩展熟龄好宅全新的价值与体验经济。

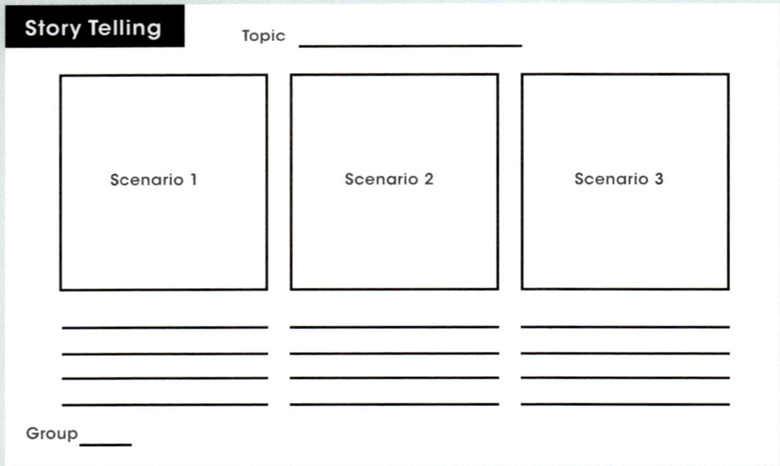

Toolkits 工具

1 Cool Hunting

Stage 1 - Scenario: the team collected domestic and global "cool" case studies of elderly care and established basic example scenarios to help sequentially understand the study. Then, the team started analyzing the users' values and behavior activities while interacting with universal design and healthcare issues to further explore potential perspectives and needs.

阶段一：情境研究。透过“酷猎”搜寻世界各地银发族酷案例，建立基础情境认知，协助参与成员逐步融入计划，探索世代的价值行为与活动，并辅以通用设计、赡养照护等相关议题，刺激出隐性观点。

2 UX Research

Stage 2 - Users: In this stage, the study team invited experts together to interview fifteen 55 to 65 year-old subjects separately, and picked out 8 selected users with distinct features for continuing research. Then, the users would share their actual living experiences and daily life habits while the team used 2D and 3D tools to present the dominant needs, potential needs, and key issues for their future living via the methods of "Contextual inquiry" and "Culture Probe."

阶段二：用户研究，通过专家与用户访谈、脉络探询、文化探针等研究方法，借由二维到三维不同工具，深入了解用户实际居住经验、日常生活习惯、潜在需求与关键议题。本阶段深入访谈 15 位 55 至 65 岁由熟龄迈向老龄的族群，最终分为 8 组进行下一阶段研究。

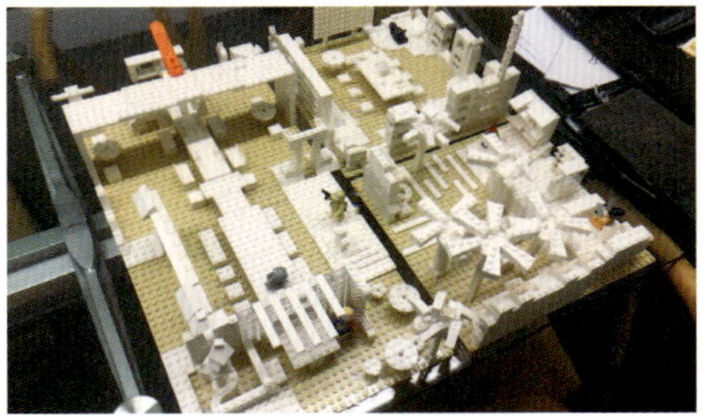

3 Co-Working

Stage 3 - Activity: Based on several meetings of a participatory workshop, the interview subjects and architects worked together on sketching out future scenarios to discover essential values and potential issues through various mock-up models and speculative scenes.

阶段三：活动共创。以参与式设计工作坊的模式，运用各种尺度的实体模型结合行为与情境，由受访者与建筑师共同草创未来场景，以勾勒出关键价值与潜在新议题为主要目标。

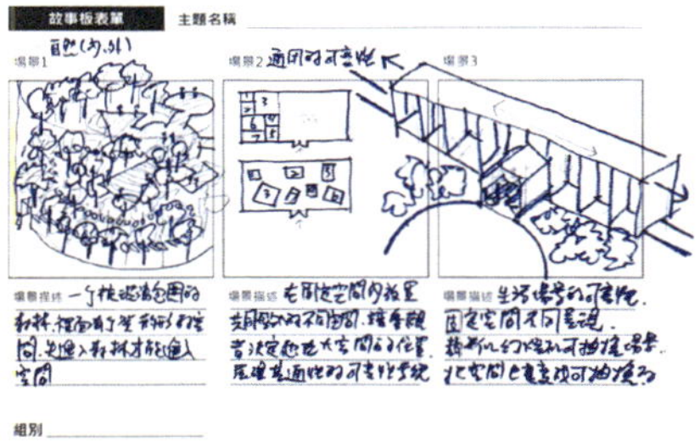

4 Prototyping

Stage 4 - Experiences: the study team made use of the results of stages 1 - 3 to shoot a schematic short film and numerous speculative scenario shots as media for discussion. Architects adopted the discussion results to draw the prototype for future development plans and buildings.

阶段四：体验共创。工作团队将前三个阶段的成果拍摄出关键分镜，以拍摄微电影及布置想象场景作为沟通媒材，协助设计者快速发展未来的开发策略及建筑原型。

Age-Friendly Cities
全龄友善城市

The study presented four future prototypes: "New Aging Dream": response to spiritual needs; "Age-Spanning Service-Oriented Housing": flexible spaces with adjustable residential units; "Leap" spaces with the latest technology devices; "Ultra Universal Design" spaces that take care of both physical and psychological needs.

In defining these four building prototypes, the study drew on the fact that the seniors aged 55 to 65 shared similar thoughts on their potential behavioral needs, calling for less private spaces and more shared spaces. The experiences and wisdom of the interview subjects may have wider applications, and therefore the team may potentially utilize the study results to develop more prototypes, extending the findings to a universal solution for all ages.

The study also shows that seniors mostly uphold a concept analogous to the "sharing-economy" and would like to adjust private space to be more "flexible" or even become public space. For instance, interviewees appreciated the idea of spaces like an Aquaponics garden, farmer's market, and beyond. Such concepts might also attract the young and middle-aged to live in the same community, subsequently developing a culturally vernacular "stacking community" and transforming it into a sustainable urban living space shared equally by seniors and youths.

The study results were transformed into an interactive exhibition in Taichung. During the exhibition, the team continued to collect opinions from the visitors. The study team hopes that more parties such as corporations, academic institutions or even government agencies will join in the research in order to develop a practical architectural and business mode, or even a true neighborhood community. The study team's ultimate goal is that the results may serve as a blueprint for sustainable "Age-Friendly Cities".

研究归纳出四种乐龄梦想宅的未来原型：回应心理要求的“熟龄梦空间”、房型单元可弹性调整的“跨龄服务宅”、结合新科技的“跃空间”以及兼顾身心需求的“超通用空间”。透过这四种原型，带出熟龄族群“小私大公”的行为特质：乐于贡献经验与智慧，随年龄演变对私有空间需求降低，转而对公共的社群空间需求增加，而这些特质延伸了单体建筑物的全龄化住宅的可能性。依照熟龄阶段的演变规律，调整建筑空间与用户活动，结合共享经济的概念，将私有空间转化为弹性空间，如住宅单元内现有空间的分享与活用；随着老化程度，进一步将弹性空间释放为公共空间，如渔菜花园、小区市集等公共领域规划；促进青壮族群加入混龄小区的可能性，并依照所在地区属性及当地特有生活文化发展“迭迭小区”，创建银青共享的可持续性城市。

计划成果以交互式展览呈现，展出过程也在持续累积建构设计参数，汇整为下一阶段的执行依据。期望透过产业、学术与政府机构合作，甚或异业联盟开发符合时代需求的商业模式，系统性地提出可执行的各式空间与邻里小区模式，进一步打造永续生活的全龄友善城市。

Exhibition flyer 展览卡片

*Note: The study defined citizens aged 50 to 65 as seniors, over 65 as elders, and 18 to 54 as young adults.

* 注：本研究计划定义熟龄族为 50–65 岁的公民；银发老龄为 66 岁以上长者；青壮年族群则为 18–54 岁之公民。

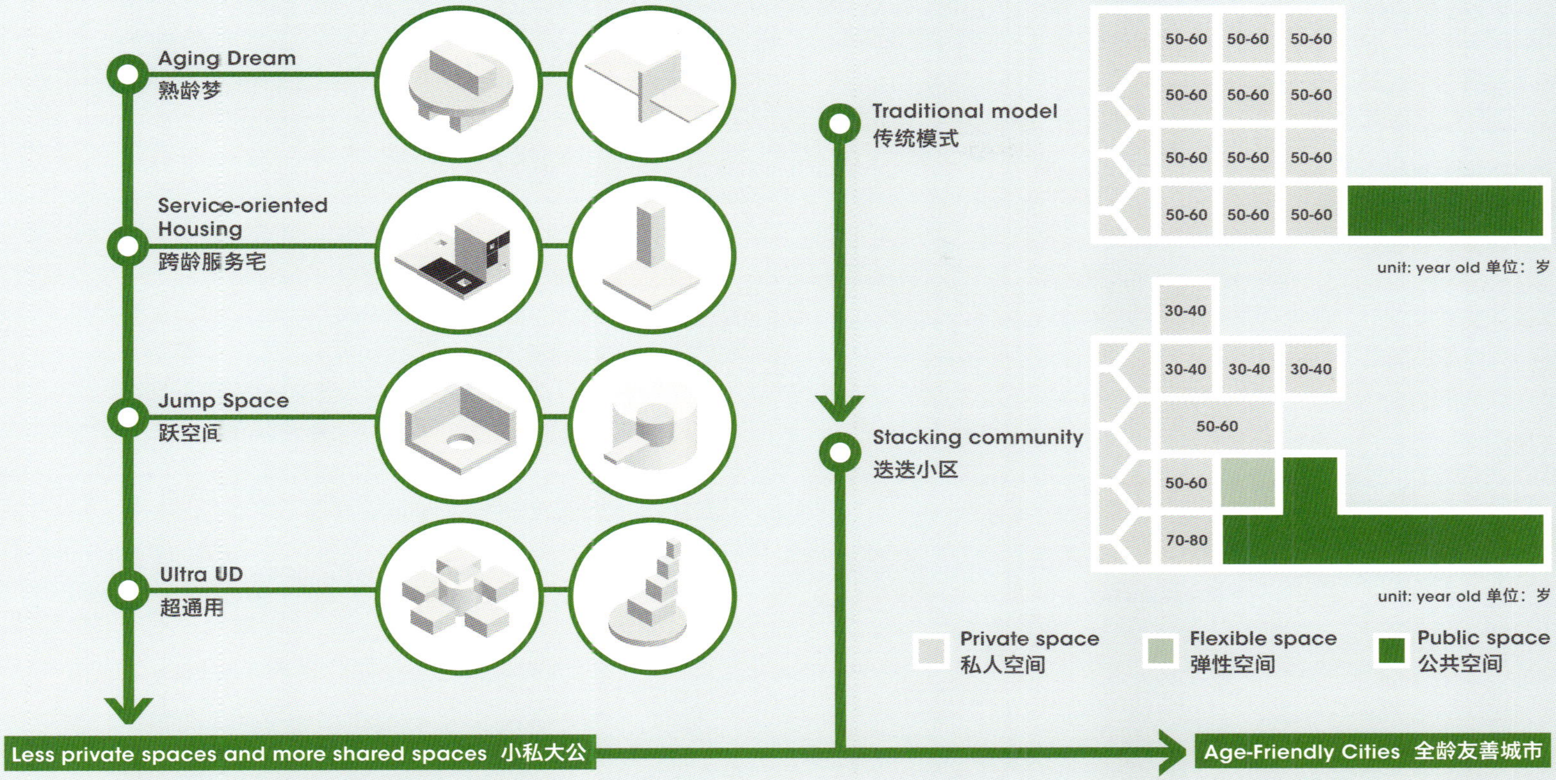
Aging Dream
熟龄梦
Service-oriented Housing
跨龄服务宅
Jump Space
跃空间
Ultra UD
超通用
Less private spaces and more shared spaces 小私大公
Traditional model
传统模式
50-60
unit: year old 单位：岁
Stacking community
迭迭小区
30-40
50-60
70-80
unit: year old 单位：岁
Private space
私人空间
Flexible space
弹性空间
Public space
公共空间
Age-Friendly Cities 全龄友善城市

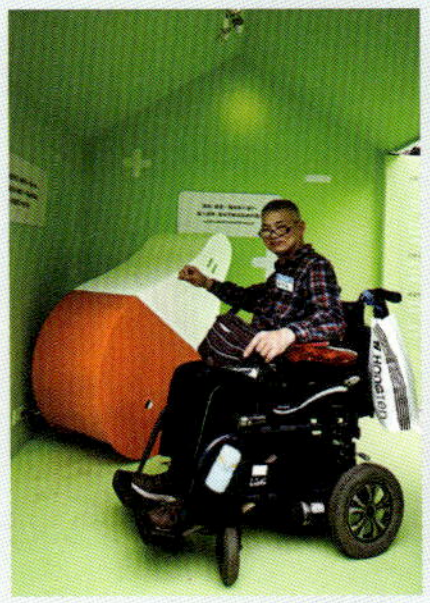

05
SHARING

Combine the resources of a large office and the flexibility of small studios to create an innovative, multi-disciplinary and collaborative working environment

具备小工作室特色的大型事务所，弹性与资源兼具的工作环境，创新、多元、共享

共享

G

An Org
Office

有机办公室

ganic

J.J. Pan and Partners' main office is located in the alleyways of a residential area in Taipei. Extending out from the residence of its founding principal, Joshua Jih Pan, the office consists of ten long, narrow row house units that are each roughly 4.5-meter-wide and 20-meter-long. Out front, a banyan tree planted by Mr. Pan has grown with the firm for more than 30 years. During this time, the firm has evolved from less than 10 people to more than 300 members, encompassing architecture, construction supervision, landscape and interior design and structural engineering, with three branch offices in Shanghai, Xiamen, and Beijing.

由潘冀带领的 JJP 团队，藏身于台北市住宅区的巷弄中，从最初一栋 4.5 米宽、20 米长的住居开始，到目前超过 10 个单元。19 号院内一株创始人手栽的橡胶树，30 多年一路与 JJP 相伴；成员自迁入时不到 10 人的编制，一路成长到工务、景观，单独另立室内设计、结构等事业体，继而跨足对岸（上海、厦门、北京）超过 300 人的规模。

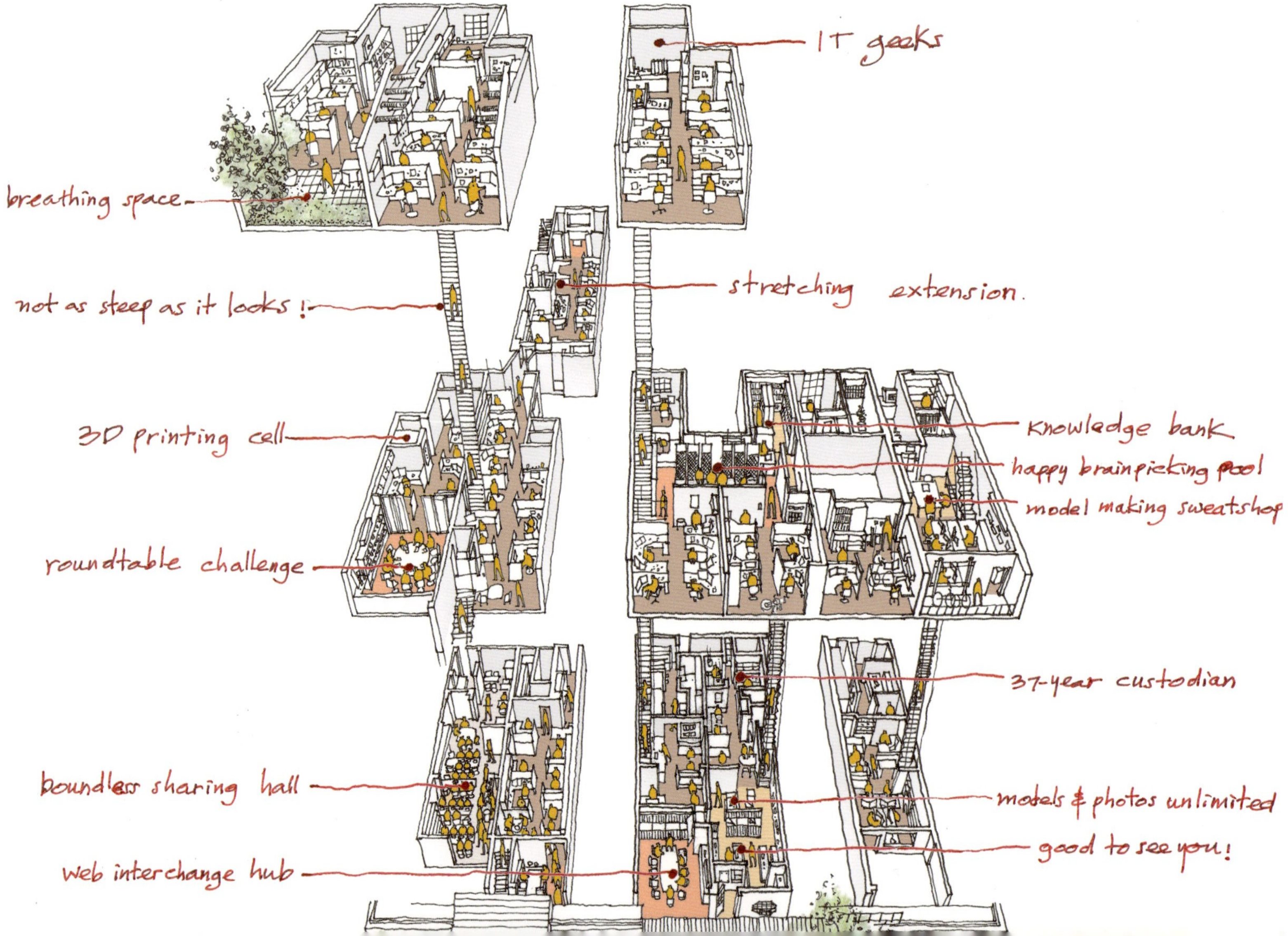
IT geeks
breathing space
stretching extension.
not as steep as it looks!
3D printing cell
knowledge bank
happy brainpicking pool
model making sweatshop
roundtable challenge
37-year custodian
boundless sharing hall
models & photos unlimited
good to see you!
web interchange hub

autumn 2017

An Organic Office

Cultivating Compassion: Humanistic Architecture as Practiced by JJP Architects and Planners

人本建筑：潘冀联合建筑师事务所的建筑实践

左列：二楼缓压平台之平行时空版 | 一点都不陡的梯 |3D 打印房 | 圆桌会议室 | 幸福食堂 / 厅堂 / 课堂 / 讲堂 | 网络无国界会议室 | 二楼缓压平台

右列：信息交换机房 | 工作区一路延伸至仁爱大道 | 知识宝山 | 脑力风暴区 | 卯起来作模型区 |37 年大家长办公室 | 无限能量模型 / 影像展示空间 | 欢喜相迎 | 屋顶花园 |40 年绿荫 | 快闪密道 | 主入口

AN ORGANIC ORIGIN

The inherent structural limitations of the row houses have made the typical open office space impossible. In its place, a group of dense, diverse and inter-dependent neighborhoods has formed. This spatial pattern also reflects, to a certain extent, the firm's work ecology-small groups, close interaction and mutual support. The library nook, model displays, and open meeting rooms are arranged in a spatial sequence to encourage design discussions and brainstorming. Furthermore, the canteen can be transformed into a conference room, a lecture hall and a training classroom through the reorganization of movable furniture.

有机发源地

JJP 办公室位于台北市静巷之中，建筑类型为始建于 1960 年代的透天厝（注：闽南地区对于联排住宅 town house/row house 的称呼）。围绕着创始人潘冀建筑师的家，随着事务所扩张逐渐蔓延为 3 层 7 栋的规模，虽然内部高密度与能量集中，但外观却异常低调、寻常。

由于结构限制无法做成开放式办公空间，却在透天厝居住格局内形成一组高密度、多样性、紧密连接的邻里单元。如此空间格局也在一定程度上反映了事务所的工作生态——小型组团、互动密切、互相支持。流线节点上安排了图书室、模型展示架与开放型会议室，鼓励设计讨论与头脑风暴的随时发生。餐厅作为一处多功能空间，通过轻型家具的移动重组，可以变身为会议室、报告厅和训练教室，举办各种学习交流分享活动。

The interior design does not adhere to a particular style. Instead, various materials and construction methods are artfully inserted into different areas of the space. For example, the fritted glass of various patterns on the exterior were originally installed to test their sun-shading and sound insulation capabilities. In order to encourage green transportation, commuting bicycles are also hung on the wall where they subtly reflect the JJP's design values and lifestyle.

空间设计并不追求材料风格的一致性，而是把各种建材样品和做法巧妙地安插在空间中，譬如各区使用不同的建筑装饰材料与照明方式，悬挂在立面外侧构件，即是为实地测试“网印玻璃”隔音遮阳效果而安装，产品示范至今近二十年，办公空间兼具了材料、工法展示的效用。为了鼓励绿色出行，通勤单车也见缝插针地安置在墙上，螺蛳壳里做道场，随着时间更替不断发生微妙变化的工作空间，反映着 JJP 的设计主张与生活方式。

DIVERSE TEAM BUILDING ACTIVITIES

The JJP bimonthly meeting tradition began in the era when Saturday was still a workday in Taiwan. The meeting is divided into two parts with the first part being an office-wide meeting, and the second part is reserved for an outside speaker. Invited speakers are not limited to architects or related professionals, so past guests have included leaders in other fields such as art and business. The aim is to broaden horizons and encourage fresh thinking and perspectives. Archi-teatime is another tradition that is a somewhat more relaxed version of the bimonthly meeting. Typically held in the office canteen, the speaker and audience can freely gather around refreshments for lively interactions. These knowledge sharing activities are the cornerstones of JJP's professionalism and dynamism. More importantly, they also testify that JJP's architectural design is not self-contained in the computer, but a vessel for real life and vision that is closely connected with society and people.

多元开拓的团队建设

团队建设对于一定规模的建筑师事务所是必不可少的，JJP 自创始以来非常重视建筑观念与技能的训练传承，而这些奠基来自日常之中。“共餐”是每日工作的节点与交流时光，用餐分梯次，定期轮换，使工作中没有交集的同仁也能互相认识；在事务所隔壁的潘冀建筑师家客厅的“读书会”，跟学院派研讨会相比更具家庭氛围与团契色彩，书目不拘一格，共餐之后通过对书籍内容观点的讨论，交流生命体悟和正向鼓励，从而调适心境应对工作中的各色境遇。“双月会”始于尚未实行周休二日的年代，逢单月择一周六上午，主要内容为工程进度汇报与嘉宾演讲，前者使事务所每个人都知道全公司所有项目的状况与进展，而非埋头在自己的工作中；特邀演讲嘉宾也不拘一格，不仅有建筑师、设计师同行贤达，也跨界到文艺、商业等领域的精英，旨在开拓眼界与思维，从不同角度获得专业上的滋养与启发。而延伸双月会嘉宾演讲的轻松版——“建筑茶会”，将地点从严肃的报告厅拉回到事务所餐厅，嘉宾与听众围坐一处，共享茶点，充分互动。这些分享活动是 JJP 专业与活力的基石，也验证 JJP 的建筑设计不是图板 / 计算机上的自娱自洽，而是与社会、与产业密切连接的真实生活与愿景的载体。

If the above-mentioned team building activities have a cross-disciplinary feel, then construction site visits are intense field educational lessons. Periodically, the firm would select several important or exciting projects under construction and arrange for all colleagues and family members to visit. Upon arrival, project designers and on-site supervisors give a presentation on the design concepts and technical characteristics. Then the group goes on a building tour together, with the senior colleagues offering analysis and conducting discussions on the spot. Such is the best professional education for passing on valuable architectural experience. It is also a great opportunity for families to witness first-hand the demands of an architectural career and enjoy the fruits of the firm's collective labor. Last by not least, the annual banquet and company trip is open to all staff members and their families. These occasions are the best opportunities for JJP to become a tight-knit family where the bonds are always close.

如果说前述团建带有跨界拓展的色彩，那么“工地参访”则是密集震撼的现场专业教育。JJP 不定期挑选若干重要或精彩的执行中的项目，安排全体同仁及家眷同往参观。负责的建筑师与现场监造会事先准备演讲稿介绍设计理念与技术特点，随后一同体验建筑实境，听主持建筑师与资深同仁分析设计得失，或就地开展讨论。这是传承建筑经验、融合现场体悟与观点交流的最佳专业教育，更是眷属们理解与认同建筑高工时及重合作特性的最好机会。
而年年广邀全员眷属参与的“员工旅游及尾牙餐宴”，便是 JJP 得以凝聚成“超级大家庭”的最佳利器，借着出游欢庆拉近一大家子的距离，情感始终紧密。

FORWARD TOGETHER

Since its inception, JJP has strived to provide a better workplace environment and advance the architectural profession. All members and partners of the JJP extended family are also encouraged to contribute their strengths towards achieving a society of shared prosperity. The firm sincerely believes in moving forward together.

Lin Ling

共进共好

长久以来，JJP 一直努力提供更好的职场环境，促进建筑师的全面成长，鼓励“大家庭”所有成员及合作伙伴们贡献所长，未来也将持续提升 JJP 影响力，带动人群、社会共同成长，“共享”美好。

凌琳

HOLLISTER
SURF BOARD

Biogra

ohies 团队成员

Joshua Jih Pan FAIA Founding Principal
潘冀 创始主持人

Joshua Jih Pan received his Master of Science degree in Architecture/Urban Design from Columbia University and Bachelor of Architecture from Rice University. He worked at the firms of, such as, Philip Johnson & Richard Foster; Davis, Brody & Associates; and Collins Uhl Hoisington Anderson in the U.S. for nearly 10 years before returning to Taiwan in 1976. His founded J. J. Pan and Partners in 1981 has since received numerous domestic and international design awards, along with scores of appearances in professional publications. He obtained China's Class-1 registered architect qualification in 2009, and has also been honored with prestigious recognitions such as Taiwan's Outstanding Architect, Taiwan Awards of Art, and Fellow of the American Institute of Architects.

美国莱斯大学建筑学士、纽约哥伦比亚大学建筑及都市设计硕士，先后于纽约菲利普 · 约翰逊和理查德 · 福斯特、戴维斯 · 布罗迪、柯林 · 悠 · 霍伊辛顿 · 安德林等建筑事务所任职。在美工作近十年后返台，于 1981 年成立潘冀建筑师事务所。作品多次发表于海内外专业建筑杂志，屡获国际建筑奖项殊荣，于 2009 年取得中国一级注册建筑师资格。其专业成就与贡献已获台湾杰出建筑师、台湾文艺奖及美国建筑师协会院士等多项终身荣誉肯定。

Jason Chen Principal
陈洁生 主持人

Jason Chen joined J. J. Pan and Partners after he earned his Master of Science in Architecture from Cheng Kung University. He firstly played as a project architect to be responsible for numerous essential project kinds including high-rises, mixed use, public housing, residential, healthcare and beyond. He then became principal afterwards, as well as the director of the mainland China. In the meantime, he magnificently implemented Knowledge Management to complement the firm's comprehensive design basis. He obtained his EMBA degree from Chengchi University in 2011 and was afforded to the 10th Outstanding Enterprise Manager Association Golden Torch award in 2013.

于台湾成功大学取得建筑硕士学位后即加入 JJP 团队，担任多项重要工程项目建筑师，范围包括高科技商办、城市综合体、集合住宅、大型健康养老地产等规划设计，2000 年成为主持建筑师，曾负责督导大陆地区开发项目，并推动知识管理平台。2011 年取得政治大学 EMBA 硕士学位，2013 年获台湾年度绩优暨潜力金炬奖经理人肯定。

Jong-Yu Cheng Principal
郑荣裕 主持人

Jong-Yu Cheng joined J. J. Pan and Partners after he received his Bachelor of Architecture from Chun Yuan Christian University. While as a Project Architect, he has been accumulating a broad range of experience of high-tech manufacturing design which emphasizes on the production efficiency and cost management. Moreover, his thorough design takes good care of the working environment for the users and the clients' needs. In 2000, he was promoted as principal and his design includes various building types: commercial, residential, institutional and public facilities.

台湾中原大学毕业后随即进入 JJP 团队工作，负责项目设计并担任高阶主管，擅长追求生产效率与成本控制的高科技厂房设计，其兼顾人性尺度与企业员工生活质量的设计思维，以及体现科技与人文并重的企图心，深获尖端科技业界信任。2000 年成为主持建筑师，作品涵盖商业、住宅、校园及公共设施等不同类型。

Chungwei Su Principal
苏重威 主持人

Chungwei Su worked for Notter Finegold + Alexander in Massachusetts, USA after receiving his Master's degree of Architecture in Urban Design from Harvard University in 1987, and became a registered architect in New York State. He returned to Taiwan in 1990 and joined J. J. Pan and Partners as a senior architect, and went to USA again in 1993 to be in charge of several multi-national design collaborations. He was promoted to Principal at J. J. Pan and Partners in 2004 and has been in charge of various planning of design innovation and organizational operation. He was awarded Taiwan Outstanding Architect in 2009 and obtained China's Class-1 registered architect qualification two years later. He has been active in architecturally academic and international events for years.

美国哈佛大学建筑硕士，美国马萨诸塞州 NF + A 建筑师事务所工作并取得建筑师执照。1990 年返台加入潘冀建筑师事务所。1993 年再度赴美负责多项跨国设计合作，2004 年担任主持建筑师，推动各项营运与创新计划；2009 年获颁台湾杰出建筑师，两年后取得中国一级注册建筑师资格。长期担任建筑设计教学、评图工作，积极参与学术评论及国际交流活动。

Chung-Tsai Huang Principal
黄种财 主持人

Chung-Tsai Huang is a graduate of Feng Chia University, where he obtained his bachelor degree of Civil Engineering. He joined J. J. Pan and Partners in 1987, assuming the position of cost estimation and site supervision. His duty included site itinerant supervising, on-site instructing, and management & execution of any other projects in the firm. In 2000, he was elected as a principal and he has greatly promoted the database of construction materials & the firm's sustainable operation.

台湾逢甲大学土木工程系毕业，台湾科技大学营建工程所研究所进修。1987 年加入事务所，负责工程预算及现场监造工作。工作项目包括现场巡回监督、指导施工、协助事务所项目管理及执行；2000 年成为主持人，积极推动所内建材、品质管理等重要资源建置与经验传承。

Robin C.K. Tang Associate Principal
唐正国 协同主持人

Robin C.K. Tang graduated from Columbia University Graduate School of Architecture, Planning and Preservation where he received his Master of Science in Advanced Architecture Design with Lucille Smyser Lowenfish Memorial Prize. He joined J. J. Pan and Partners in 1994, promoted through the positions of project architect, project manager, and associate of the design department. He has been staying in the Mainland China, mainly Shanghai & Xiamen after 2002. A position of Adjunct Professor in School of Architecture and Civil Engineering in Xiamen University came to him since 2012. He has been an associate partner and one of the heads of the mainland China area to lead the planning and designs of the three offices.

美国纽约哥伦比亚大学建筑设计硕士，曾获露西尔 · 斯迈泽 · 罗文费雪纪念奖学金 。返台后于 1994 年加入 JJP 团队，担任设计部项目经理、协理等职务，2002 年后常驻上海、厦门两地，并于 2012 年起担任厦门大学建筑学院兼职教授。2015 年成为协同主持人，负责大陆地区之规划及设计项目，现任 JJP 大中华区共同负责人之一。

Hsiao-Ming Chang Associate Principal
张晓鸣 协同主持人

Hsiao-Ming Chang received his Master of Science in Architecture from University of Pennsylvania. He has registered his architect profession in California and acquired LEED® Accredited Design Professional. Before moving back to Taiwan, he had accumulated abundant experiences in designing educational institutions. He then joined J. J. Pan and Partners in 2005 as project manager to lead the team of advanced research facilities and medical projects. He became associate partner in 2014 and has been in charge of integrating administrative resources & personnel incubator program.

美国宾夕法尼亚大学建筑硕士，美国注册建筑师及 LEED 专业认证成员。在美期间执行多项校园系馆及相关设施等设计规划项目。2005 年返台后加入 JJP 团队担任项目经理，带领群组参与先进设备、特殊疗程等医疗项目执行。2014 年成为协同主持人，协助统合行政资源及各项人事培养计划。

Ya-Tian Chuang Associate Principal
庄雅典 协同主持人

He obtained his Master degree of Architecture from Washington University in St. Louis and Employed Master of Business Administration degree from Guanghua School of Management, Peking University. With more than 30-year accumulated experience, his practices have been across a broad range in commercial field such as: property, theme park, HQ, hospitality, residential and MRT system. He has taken leads in development, planning, design & operation of large-scale office projects. He is now one of the heads of the mainland China area to lead the planning and designs of the three offices.

美国注册建筑师，美国圣路易市华盛顿大学建筑硕士，北京大学光华管理学院EMBA硕士。拥有超过30年设计经验，作品包括商业地产、主题公园、企业总部、旅馆、住宅、捷运系统等，擅长商业地产开发与设计，具有大型商办项目的开发管理、整体规划建筑设计等实务经验，2016 年加入成为 JJP 大中华区共同负责人之一。

Chiu-Hwa Wang Consulting Architect
王秋华 顾问建筑师

Chiu-Hwa Wang received her degree of Master of Science in Architecture from Columbia University. She joined Percival Goodman in 1950, obtained architectural license and became a partner in 1960. In 1979, she moved back to Taiwan and began teaching in a number of institutions. Her design experiences accumulated from numerous cultural institutions and campus planning projects have earned her a renowned reputation. She has collaborated with J. J. Pan and Partners on more than 40 projects, including over 20 libraries. She is now an architectural consultant at J. J. Pan and Partners.

美国哥伦比亚大学建筑及城市设计硕士。1950 年进入普西沃 · 古德曼事务所工作 ,1960 年取得美国建筑师执照后，成为该事务所协理、合伙人，直至 1979 年古德曼退休。是年返台定居、任教、成立事务所，并与 JJP 团队展开长期合作伙伴关系。作品以文化研究设施及校园规划见长，尤以图书馆作品备受赞誉，与 JJP 团队共同完成 40 余件作品中，大学图书馆即达半数以上，现任 JJP 团队顾问建筑师。

Associates
协同负责人

赵化宇 Chao, Chris Huayu
黄巾倪 Huang, Genie
黄介帅 Huang, Mitchell C. S.
廖纪勳 Liao, Chi-Hsun
陈伟民 Sae Tang, Oui Ming

张志明 Chang, Chih-Ming
张国兴 Chang, Kao-Hsin
张元琳 Chang, Mallet
陈昭邑 Chen, Chao-Yi
陈绍平 Chen, Shao-Ping
陈伯桢 Chen, Steven B. J.
郑仁杰 Cheng, Jen-Chieh
郑顺益 Cheng, Shun-Yi
覃亦仙 Chin, Yi-Shien
贺照梅 Heh, Grace
侯秉宜 Hou, Pin-I
谢宗芳 Hsieh, Kevin F.
许佳阳 Hsu, Chia-Yang
许　国 Hsu, Steve Kao
黄诗芳 Huang, Shih-Fang
黄彦霖 Huang, Yen-Lin
洪宏达 Hung, Hung-Ta
洪亮辰 Hung, Liang-Chen
李春生 Lee, Chun-Sheng
李大威 Lee, David
李　翩 Lee, Pen
李昱霖 Lee, Yu-Lin
梁建元 Liang, Chien-Yuan
廖志坚 Liao, Frank C. C.
廖珮雯 Liao, Pai-Wen
林镱汰 Lin, Simon Y. T.
毛以励 Mao, I-Li
沈庆豊 Shen, Ching-Feng
苏怡樵 Su, Yi-Chyau
蔡恒升 Tsai, Ady
蔡政颖 Tsai, Cheng-Ying
蔡志伟 Tsai, Chih-Wei
蔡祖烈 Tsai, Tzuu-Lieh
曾彦智 Tseng, Yen-Chih
王承恩 Wang, Cheng-En
王秋斌 Wang, Chiu-Ping
王文錤 Wang, Vincent
王文棋 Wang, Wen-Chi
翁源彬 Weng, Yuan-Pin
吴方诚 Wu, Fang-Cheng
吴轩铭 Wu, Hsuan-Ming
杨念恒 Yang, Nien-Heng
杨天阖 Yang, Tien-Kai
叶胜钿 Yeh, Sheng-Tien
严纬勇 Yen, Henry W. Y.
周炳杉 Zhou, Bing-Shan

Awards and Honors

2018 Outstanding Intelligent Building Design Award
tsmc Fab 14 Phase 5 Office

Outstanding Award, FIABCI-Taiwan Real Estate Excellence Award
Ally Logistics Yangmei Center

2017 Quality Award for Public Construction, Taiwan Gold Award for Architecture
Taichung Metro Green Line – DG02

Outstanding Green Building Design Award
tsmc Fab 14 Phase 5 Office

Special Award, New Taipei City Urban Design Award
Composite Municipal Hospital

Honorable Mention, Center of the Built Environment's Livable Building Award
Delta Americas Headquarters

Shortlisted, Leading European Architecture Forum Award – Refurbishment
Taichung First High School Heritage Building

BUILD-News - Most Anticipated Taiwan Mixed-Use Project
TVBS Media Park

2016 Gold Award, WACA Housing and Residential Design Award
Xinglong Public Housing Block 1

2015 Shortlisted, WAN Future Projects Civic Award
New Taipei Museum of Art

Outstanding Award, Public Construction Golden Quality Award for Architectural Design
Xinglong Public Housing Block 1

Outstanding Green Building Design Award
tsmc Fab15 Phase 1 Office

Taipei City Distinguished Public Construction Award
Xinglong Public Housing Block 1

Outstanding Award, FIABCI-Taiwan Real Estate Excellence Award
KUAS Library & Computer Center

Jury Awards+Popular Choice Awards, Architizer A+ Awards (Pop-up & Temporary)
Ring of Celestial Bliss

2014 Gold Award, WACA Housing and Residential Design Award
Urban Renewal Project, Ren-Ai/Jiang-Kuo Thoroughfare

Good Design Award, WACA Housing and Residential Design Award
Ministry of Finance Joint Development Project – Huaku New World

The 3rd Taichung Urban Design Awards
Library of Public Information

2013 Outstanding Award, FIABCI-Taiwan Real Estate Excellence Award
Advantech Smart Campus Phase 1

Outstanding Green Building Design Award
Plant Phase II, Delta Electronics, Inc.

Good Design Award, WACA Awards
Library of Public Information

1st Prize, CSR Award for Environment Protection, *Global Views* Monthly
Ring of Celestial Bliss

2012 Gold Award, The International Design Awards for Architecture
Library of Public Information

Innovative Design of the Year, The 10th Golden Torch Awards
Library of Public Information

First Prize in Construction Quality, the 14th Golden Award for Architecture
Chocolate Museum

Golden Quality Award, FIABCI-Taiwan Real Estate Excellence Award
Chocolate Museum

2011 Outstanding Green Building Design Award
tsmc Fab12 phase 4

2010 Public Construction Excellence Award for Accessibility
Suang-Lien Senior Living Community

Public Construction Golden Quality Award for Architecture
Kinman County Datong Home

2009 Excellence in Innovative High-tech Construction Award, Science Park
tsmc Fab12 Phase4, connection bridge

Jiangxi Outstanding Architecture Design Award, 2nd place
Nanchang Factory Plant

Quality Award, FIABCI-Taiwan Real Estate Excellence Award
Scientific Research Building, Feng Chia University

2008 Gold Plaque Award, Taoyuan Architecture Award for Industrial/Commercial Building
TRI Headquarters

Taiwan Interior Design Award of Working Space
Kingyorker Headquarters Building

AIA New York State Award of Excellence for Adaptive Reuse
Horizon Design Shanghai Office

Best Historic Preservation Project, The 2nd Biennial *Business Week / Architectural Record* China Awards
Horizon Design Shanghai Office

The Luban Prize for Construction Project
SMIC Wuhan Office

Chutian Cup for Hubei Architectural Quality Award
SMIC Wuhan Office

Hubei Structure Quality Award
SMIC Wuhan Office

2006 Citation, *Taiwan Architect* Annual Design Award
Scooter Garage, Hsinchu Chiao Tung University

Silver Plaque Award, Taipei City New Look Award for Open Space
Taipei Truth Lutheran Church

AIA New York State Citation Award
Taipei Truth Lutheran Church

Shanghai Outstanding Residential Project Award, Monomial Design, 3rd place
Greenland Shanghai Plaza

2005 Gold Medal Award, WACA Awards
Scooter Garage, Hsinchu Chiao Tung University

Silver Plaque Award, Taoyuan Architecture Award for Housing
Acer Long Tern Residential Development III

Silver Plaque Award, Taoyuan Architecture Award for Industrial Building
Quanta TFT/LCD Plant

2004 Public Construction Golden Quality Award for Architectural Design
Szu Zhih New Town

2003 China Service Apartment Design Prize, 2nd place
Greenland Shanghai Plaza

2001 AIA New York Chapter Annual Design Award
Da Xia House, Chinese Culture University

Taiwan Building Design Award for Intelligent Building
Macronix Headquarters Building

1999 Renowned Finalist, Far Eastern Outstanding Architects Design Award
Founder's Memorial Library, Chinese Culture University

Quality Excellence Award, Taiwan Public Construction Annual Award
National Museum of Marine Biology/Aquarium

1998 Taiwan Building Design Award
Engineering Building, Chung Yuan Christian University

1997 Taiwan Building Design Award for Passive Energy Efficiency
Library, Chung Yuan Christian University

Architecture Golden Award
My House Cherry Grove Housing

1995 Taiwan Building Design Award
Student Center, Hsinchu Chiao Tung University

Architecture Golden Award
Synpac Sunshine State Housing

1994 Citation, *Taiwan Architect* Annual Design Award
Main Library & Information Sciences Center, Chung Cheng University

1993 Silver Plaque Award, *Taiwan Architect* Annual Design Award
Sung Chiang Poetry Park

Taiwan Building Design Award
Main Library, Changhua University of Education

1992 Silver Plaque Award, *Taiwan Architect* Annual Design Award
Gymnasium and Indoor Swimming Pool, Chung Yuan Christian University

1991 Taiwan Building Design Award for Passive Energy Efficiency
Precision Instrument Development Center for the National Science Council

Citation, *Taiwan Architect* Annual Design Award
Medical Research Building & Conference Center, Taichung Veterans General Hospital

1990 Taiwan Building Design Award
Medical Research Building & Conference Center, Taichung Veterans General Hospital

1989 Taipei Municipal Building Design Award
Overseas Radio and Television New Office

1988 Taiwan Building Design Award
Library, Chung Yuan Christian University

Ten Most Outstanding 'Creativity in Architecture' Award
Gymnasium, Taiwan Ocean University

1985 Gold Plaque Award, *Taiwan Architect* Annual Design Award
Library, Chung Yuan Christian University

奖项

2018 台湾优良智慧建筑奖
台湾积体电路南科十四厂区五期办公栋

台湾卓越建设奖最佳规划设计类卓越奖
永联杨梅物流中心

2017 台湾建筑金奖公共建设优质奖（规划设计类）
台中捷运绿线 DG02 标

台湾优良绿建筑设计奖
台湾积体电路南科十四厂区五期办公栋

第三届新北市都市设计大赏（规划设计类）特别奖
新北市立土城医院

美国 CBE 宜居建筑奖佳作
台达美洲总部

英国 LEAF 建筑再利用奖入围
台中一中校史馆修复及再利用

英国 BUILD-News 台湾最受瞩目商业综合体奖
TVBS 媒体园区

2016 世界华人建筑师协会住宅与住区设计金奖
兴隆公共住宅一区

2015 WAN 未来城市奖—城市奖入围
新北市立美术馆竞图

公共工程金质奖建筑设计特优奖
兴隆公共住宅一区

台湾优良绿建筑设计奖
台湾积体电路中科十五厂区一期办公栋

台北市政府公共工程卓越奖—建筑类第一级
兴隆公共住宅一区

台湾卓越建设奖最佳规划设计公共建设类卓越奖
高雄应用科技大学图书资讯大楼

Architizer A+ 瞬间快闪建筑评审奖 / 观众票选奖
永续之环

2014 世界华人建筑师协会住宅与住区设计金奖
台北市仁爱建国都市更新案

世界华人建筑师协会住宅与住区设计—居住综合体优秀设计奖
财政主管部门二地合作开发项目—华固新天地

第三届台中市都市空间设计大奖优胜
公共资讯图书馆

台湾卓越建设奖—最佳施工质量卓越奖
研华智能园区一期

2013 台湾优良绿建筑设计奖
台达台南二期厂房

世界华人建筑师协会优秀奖
公共资讯图书馆

《远见》杂志“企业社会责任奖”环境保护组首奖
永续之环

2012 IDA 国际设计建筑金奖
公共资讯图书馆

第十四届国家建筑金质奖—施工质量类首奖
巧克力博物馆

台湾卓越建设奖—最佳施工质量金质奖
巧克力博物馆

年度创新设计 / 第十届金炬奖
公共资讯图书馆

2011 台湾优良绿建筑设计奖
台湾积体电路竹科 12 厂区四期

公共建筑物无障碍选拔优良建筑物优等奖
双连教会社会福利园区

2010 公共工程金质奖建筑设计优等奖
金门县大同之家

2009 科学工业园区科技建筑服务创意奖
台湾积体电路竹科十二厂区四期连通管桥

江西省优秀工程勘察设计行业二等奖
南昌菱光科技一期工程 T 栋办公楼

台湾卓越建设奖最佳施工质量类优质奖
逢甲大学科研大楼

2008 桃园县桃花源建筑金质奖
德律科技林口厂办大楼

台湾室内工作空间类设计奖
勤裕企业办公大楼

美国纽约州建筑师协会建筑再利用杰出奖
群裕设计咨询上海办公室

第二届《美国商业周刊》暨《建筑实录》中国奖（最佳历史保护）
群裕设计咨询上海办公室

中国建筑工程鲁班奖
武汉新芯集成电路生产项目办公楼

湖北省建筑优质工程奖（楚天杯奖）
武汉新芯集成电路生产项目办公楼

湖北省建筑结构优质工程
武汉新芯集成电路生产项目办公楼

2006 台湾《建筑师》杂志台湾建筑佳作奖
新竹交通大学机车停车棚

台北都市彩妆征选开放空间改善银奖
台北真理堂

美国纽约州建筑师协会佳作奖
台北真理堂

上海市优秀住宅工程单体设计元素三等奖
上海延安西路绿地广场

2005 世界华人建筑师协会金奖
新竹交通大学机车停车棚

桃园县桃花源建筑奖住宅类银奖
龙腾渴望园区

桃园县桃花源建筑奖厂房类银奖
广辉电子液晶薄膜显示器厂房

2004 公共工程金质奖建筑设计特优奖
台北县四知八村

2003 中国酒店式公寓建设整体评比全国第二名
上海延安西路绿地广场

2001 美国纽约市建筑师协会年度建筑设计奖
中国文化大学城区部大夏馆增建

台湾建筑研究所优良智能型建筑奨
旺宏电子总部大楼

1999 第一届远东杰出建筑设计入围奖
中国文化大学创办人纪念图书馆

台湾公共工程质量优良奖
垦丁海洋生物博物馆第一期

1998 台湾优良建筑设计奖
中原大学工学馆

1997 台湾建筑节约能源优良设计奖
中原大学张静愚纪念图书馆

台湾建筑金奖
美好林口樱花山庄

1995 台湾优良建筑设计奖
新竹交通大学活动中心

台湾建筑金奖
欣业永阳光加州一、二期

1994 台湾《建筑师》杂志佳作奖
中正大学图书暨资讯大楼

1993 台湾《建筑师》杂志银牌奖
松江诗园

台湾优良建筑设计奖
彰化师范大学总图书馆

1992 台湾《建筑师》杂志银牌奖
中原大学体育馆游泳池

1991 台湾建筑节约能源优良设计奖
国科会精密仪器中心

台湾《建筑师》杂志佳作奖
台中荣总医学研究大楼及会议厅

1990 台湾优良建筑设计奖
台中荣总医学研究大楼及会议厅

1989 台北市政府优良建筑设计奖
救世传播协会办公大楼

1988 台湾优良建筑设计奖
中原大学张静愚纪念图书馆

台湾第一届十大杰出创意建筑师金塔奖
海洋大学育乐馆

1985 台湾《建筑师》杂志金牌奖
中原大学张静愚纪念图书馆

TIMELINE

01

02

03

1981

▶ J.J. Pan Architects founded with a staff of 3
成立潘冀建筑师事务所，成员三人

▶ (01) NTOU Multiple-Stage Renewal, Keelung
海洋大学校园规划扩建

▶ China Grain Products R&D Institute, Taipei
中华谷类食品研究所

▶ CYCU Chang Ching Yu Memorial Library, Chungli
中原大学张静愚纪念图书馆

1983

▶ Joshua Pan became managing director of *Taiwan Architect* Monthly, served until 1986
潘冀任台湾建筑师杂志社社长至1986 年

1986

▶ (02) NTOU Main Library, Keelung
海洋大学图书馆

▶ Medical Research Building & Conference Center, Taichung Veterans General Hospital, Taichung
台中荣总医学研究大楼及会议厅

1991

▶ J.J. Pan Architects' 10th Anniversary Commemoration Exhibition held in the office
十周年作品展—仁爱路办公室

▶ Sung Chiang Poetry Park, Taipei
松江诗园

▶ CYCU Indoor Swimming Pool, Chungli
中原大学室内游泳池

▶ (03) NCTU Student Center, Hsinchu
新竹交通大学学生活动中心

▶ (04) Winbond IC Wafer Fab II, Hsinchu
华邦科技竹科二厂

VIRTUAL | ARCHITECTURE

1983

▶ CYCU Civil Engineering Building, Chungli
中原大学水土馆竞赛

A

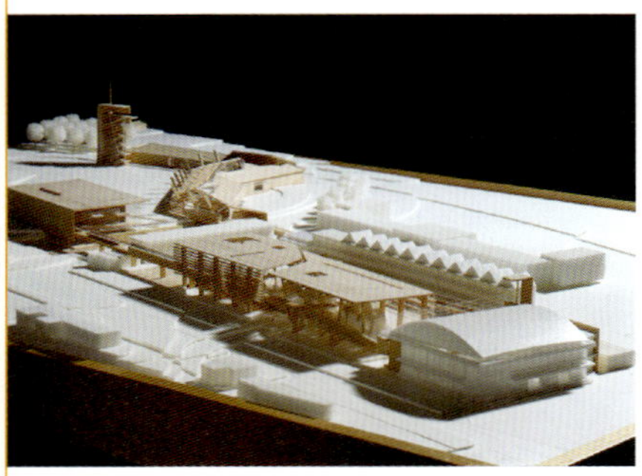

B

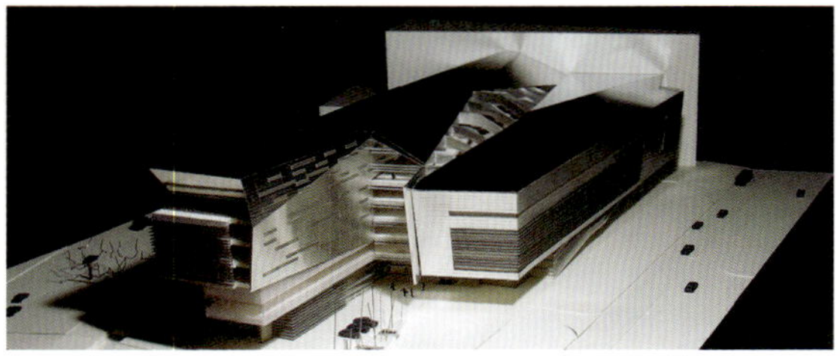

C

04

05

06

07

08

▶ Joshua Pan elected as Fellow of the American Institute of Architects
潘冀获选台湾首位美国建筑师协会院士

▶ Joshua Pan elected as Taiv
Outstanding Architect
潘冀获选台湾杰出建筑师

1994

1996

▶ (05) NHHU Information Science & Electrical Engineering Building, Hsinchu
新竹清华大学资讯电机大楼

▶ NCCU Main Library & Information Sciences Center, Chiayi
中正大学图书资讯中心

▶ Yung Ho Church, Taipei
永和教会

▶ NTOU Engineering College Building No.1, Keelung
海洋大学工学院一馆

▶ (06) Shichih R&D Center, Development Center for Bio-Technology, Taipei
汐止生技研发中心

▶ My House Cherry Grov
Housing, Taipei
美好建设林口樱花山庄

▶ tsmc Fab 3 & 4, Hsinch
台积电竹科三四厂

▶ Synpac Sunshine State
Housing, Taoyuan
欣业永阳光加州一、二期

▶ Grace Gospel Center, Mon-Eng Presbyterian Church, Taichung
蒙恩教会

D

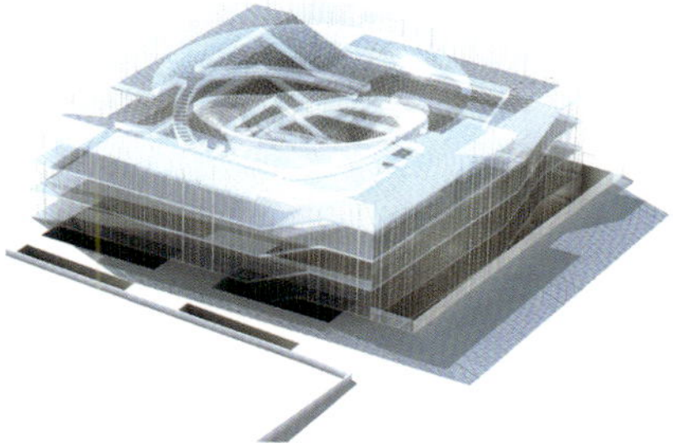

E

23

24

25

h Anniversary
emoration Student
etition & Forum, Tianjin
年活动，“海峡两岸大学生建
竞赛暨建筑论坛”，天津

▶ *Brick by Brick - Building the Foundation of Life by Joshua Pan* is published
《人生基本功：建筑师潘冀的砌砖哲学》发行

▶ Third monograph titled *Master Architect Series – J.J. Pan and Partners*, is published
第三本英文作品专书发行

▶ Joshua Pan elected as Outstanding Alumni of Cheng Kong University
潘冀获颁成功大学校友杰出成就奖

▶ Jason Chen elected as Outstanding Enterprise Manager Association Golden Torch Merits Award
陈洁生获颁绩优暨潜力金炬奖经理人

1

2012

2013

ry of Public Information,
ung
资讯图书馆
colate Museum Taoyuan
力博物馆
tgers Community
tian Church, NJ, USA
纽泽西州若歌教会
U International
e, Taipei
大学第二国际学生宿舍
a Tropical Resort

Hotel, Hainan, China
海南岛三亚酒店

▶ NCCU Research and Incubation Center, Taipei
政治大学研究暨创新育成总中心

▶ NTUH Cancer/Proton Center, Taipei
台大医学院癌症中心医院

▶ Delta Tainan Plant Phase II
台达台南二期厂房

▶ VIA Shenzhen R&D Center
威盛深圳创新设计中心

▶ Dingpu MRT Station, Taipei
顶埔捷运车站

▶ Taiwan Photon Source Facility, Hsinchu
台湾光子源加速器实验中心

▶ Ring of Celestial Bliss, Hsinchu
永续之环

▶ tsmc Fab15 Phase 1, Taichung
台积电中科十五厂一期

▶ Bethany Children's Home, Taipei
伯大尼少儿家园

▶ Advantech Smart Campus Taipei
研华智能园区

▶ KASTU Library & Computer Center, Kaohsiung
高雄应用科技大学图书资讯大楼

▶ tsmc Fab14 Phase 5, Tainan
台积电南科十四厂五期

▶ NTUH Radiation and Proton Therapy Center, Tainan
台大医学院辐射科学暨质子治疗中心

▶ Daxi Sports Court Shelter, Taitung
大溪小学风雨球场

▶ Giant Bicycles Headquarters, Taichung
巨大机械总部大楼

▶ Delta Americas Headquarters, CA, USA
台达美洲总部

▶ Suang-Lien Xinzhuang Social Welfare Center, Taipei
双连新庄社会福利中心

▶ Xinglong Public Housing Block 1, Taipei
兴隆公共住宅一区

▶ Xinheng Taipei Neihu Office Building
欣恒内湖办公大楼

2012

2013

2014

ship Building, Huashan
l Creative Center, Taipei
化创园区旗舰店

▶ M TVBS Media Center, Taipei
TVBS 媒体园区

▶ N BAIC Group R&D Center
北京汽车（零部件）研发中心

▶ Beijing Agricultural Products Logistics Center
北京农产品物流园

▶ O New Taipei Museum of the Art
新北市立美术馆

▶ P Tittot Taicang Artist Workshop Jiangsu, China
琉园（太仓）艺术家工作室

▶ Haicang Dapingge (Landscape) Promenade, Xiamen
海沧大屏阁观景区

▶ Loop House, Shenzhen, Guangdong, China
深圳正中别墅

O

P

09

10

11

12

an

▶ Joshua Pan selected as 200 Most Influential Personalities of the Millennium by *Commonwealth* Monthly

潘冀获选入《天下杂志》“影响 200，飞越 2000”200 位人物

▶ First monograph titled *Master Architect IV* – J.J. Pan and Partners, is published. First architectural monograph published in English in Taiwan

台湾第一本全英文建筑作品专书发行

▶ Tran
Part
潘冀
成员

1998

1999

200

e

u

▶ Etron Headquarters Building, Hsinchu
钰创科技大楼

▶ (07) CCU Founder's Memorial Library, Taipei
中国文化大学创办人纪念图书馆

▶ CYCU Engineering Building, Chungli
中原大学工学馆

▶ Neihu Feng Chi Housing, Taipei
丰麒内湖住宅

▶ (08) Primax Headquarters Building, Hsinchu
致伸总部大楼

▶ Macronix Headquarters Building, Hsinchu
旺宏总部大楼

▶ (09) CCU Da Xia House Addition and Renovation, Taipei
中国文化大学城区部大夏馆增建

▶ (10) Acer Long Tern Residential Development III, Taoyuan
宏基龙潭渴望园区

▶ Museum of Marine Biology & Aquarium, Pingtong
屏东海洋生物博物馆第一期

▶ Hsinchu Standard Factory Stage One Reconstruction
竹科标准厂房改建一期

F

G

20

21

22

▶Chungwei Su elected as Taiwan Outstanding Architect
苏重威获选台湾杰出建筑师

▶Joshua Pan awarded Taiwan Real Estate Excellence Award for Special Contribution towards Development and Property Construction, by FIABCI-TW
潘冀获颁台湾建设特别贡献奖

▶Joshua Pan elected as Honorary Professor of Jiangnan University, China
潘冀获任中国江南大学荣誉教授

▶JJP 30
Comm
Comp
三十周
筑设计

2007

- Huga Fab III and Headquarters, Taichung
  广镓中科三厂暨总部大楼
- (21) Lien Hwa Headquarters, Taipei
  联华南港总部大楼
- Pegatron Shanghai R&D Center
  和硕上海研发中心
- Creative Sensor Nanchang Factory Plant Phase I - Building T, Jiangxi,China
  江西南昌菱光厂房一期 T 栋办公楼
- Neo Solar Power Headquarters, Hsinchu
  新日光能源总部大楼
- tsmc Fab12 Phase 4, Hsinchu
  台积电竹科十二厂四期
- Church of the Suang-Lien Senior Living Community, Taipei
  双连教会社会福利园区礼拜堂

2009

- Mackay Medical College, Taipei
  马偕医学院校舍
- Taoyuan intl' Airport MRT Stations A22, A23
  台湾桃园国际机场捷运车站 A22, A23

201

- Libra
  Taic
  公共
- Cho
  巧克
- (22) R
  Chri
  美国
- NCC
  Hou
  政治
- San

2007

▶E Jianfa Changsha Real Estate Planning, Hunan, China
建发长沙住宅项目

▶F Ningbo Ind Design & Creative Center Project, Zhejiang, China
宁波工业创业设计园区

▶G Hua Nan Bank HQ, Taipei
华南银行总行世贸大楼

▶Xiamen Supervising Building China
厦门监管大楼竞赛

2009

▶H Taiwan Dramatic Arts Center Project, Taipei
台湾戏剧中心

▶I hTC R&D Project, Taipei
宏达电子总部大楼

▶J Pingtung Performing Concert Hall
屏东演艺厅

2010

▶K Urban Renewal Project, Ren-Ai/ Jiang-Kuo Thoroughfare, Taipei
台北市仁爱建国都市更新项目

201

▶L Flag
Cultur
华山文

M

N

13

14

15

…oned to J.J. Pan and
…rs, with a staff of over 150
…筑师事务所改制联合，
…破 150 人

▶ JJP 20th Anniversary Commemoration Concert held at the NTCH in Taipei
二十周年活动 “音乐、建筑与环境” 多媒体音乐会 - 台北两厅院

▶ Horizon Design, JJP's Shanghai office, founded
群裕设计咨询（上海）公司成立

▶ Joshua Pan elected as Fellow of the Architectural Institute of Taiwan
潘冀获选台湾建筑学会会士

2001

- ⑪ Quanta TFT/LCD Plant, Taoyuan
  广辉电子液晶薄膜显示器厂房
- ⑫ Nan Kai College Administration Building, Nanto
  南开技术学院行政大楼
- Winbond Technology Building, Taipei
  华邦科技大楼
- ⑬ Macronix Recreation/Training Center, Hsinchu
  旺宏员工训练中心
- Suang-Lien Senior Living Community, Taipei
  双连教会社会福利园区
- tsmc Fab 12 & Headquarters, Hsinchu
  台积电竹科十二厂暨总部大楼
- ⑭ ASUSTeK Headquarters Complex, Taipei
  华硕电子关渡厂区

2003

- BJSC Fab IV & V, Beijing
  北京中芯晶圆厂
- ⑮ NCTU Scooter Garage, Hsinchu
  新竹交通大学机车停车棚
- CYCU Holistic Education Village, Chungli
  中原大学全人教育村
- Kingland Mansion, Shanghai
  上海金隆置地住宅大楼
- ZyXEL R&D Campus in Wuxi, Jiangsu, China
  合勤无锡研发中心
- Taipei Truth Lutheran Church
  台北真理堂
- FCU Science and Aeronautical Engineering Building, Taichung
  逢甲大学科学与航太馆
- ⑯ CCU Gymnasium, Taipei
  中国文化大学体育馆

2001

- Ⓐ 921 Earthquake Museum, Nanto
  921 地震博物馆

2003

- Shengming Zhiyuan International Plaza, Beijing
  北京生命之源国际竞赛
- Ⓑ Quanta R&D Center Project, Taipei
  广辉电子研发中心

2004

- Fiume Veneto, Italy
  意大利小区改建竞赛
- Ⓒ Xinyi A12 Hotel Development, Taipei
  信义区 A12 住商混合大楼开发

H

I

16

17

18

1

▶ Second monograph titled, *FREEDOM IN DESIGN*, is published.
第二本英文作品专书发行

▶ Horizon Design opened Xiamen Office
群道设计咨询（厦门）公司成立

2006

▶ 17 MediaTek Headquarters, Hsinchu
联发科技竹科总部大楼

▶ Shanghai Greenland Plaza
上海延安西路绿地广场

▶ Horizon Design Shanghai Office
群裕设计咨询上海办公室

▶ FJCU Research Building, College of Medicine, Taipei
辅仁大学医学综合大楼

▶ 18 Fleur de Chine Hotel, Sun Moon Lake, Nanto
云品日月潭酒店

▶ Taipei Street Furniture Design
台北街道家具设计

▶ ZyXEL Recreation Center, Hsinchu
合勤竹科活动中心

▶ 19 FCU Xue Si Building, Taichung
逢甲大学学思楼

▶ Kingland Ningbo Financial Center, Zhejiang, China
金隆宁波商务中心

▶ Kingyorker Headquarters, Taipei
勤裕企业总部大楼

▶ TRI Headquarters, Taoyuan
德律科技研发总部

▶ 20 IMPAX Lab Taiwan Plant Phase I, Chunan
益邦制药竹南一期厂房

▶ High Tech Computer Nanhui Plant, Shangh
威宏南汇厂区

▶ Ginling Girls' High Sch
Buildings, Taipei
金陵女中校舍

2005

▶ Da Hua Linshui Conceptual Planning, Dalien,China
大华集团大连项目竞赛

2006

▶ D Stockholm Library, Danmark
斯德哥尔摩图书馆竞赛

J

K

L

TIMELINE

26

27

28

29

2015

- Joshua Pan awarded Taiwan Awards of Art
  潘冀获颁台湾文艺奖

2016

- JJP Shanghai founded
  上海潘冀成立
- 23 NTNU Linko Campus Information and Teaching Building, Taipei
  师范大学林口校区资讯与教学大楼
- 24 Cathay Life Insurance Neihu Office Building, Taipei
  国泰人寿内湖潭美办公大楼
- Taichung First High School Heritage Building
  台中一中校史馆修复及再利用
- Campus Evangelical Fellowship Building, Taipei
  校园福音团契大楼
- NCCU Center for Public and Business Administration Education, Taipei
  政治大学公共行政及企业管理教育中心
- Taoyuan intl' Airport MRT Stations A1 to A6, Taipei
  台湾桃园国际机场捷运车站 A1-A6
- 25 E. Sun Bank Building Complex, Taipei
  玉山银行信用卡大楼

2017

- JJP Beijing founded
  北京潘冀成立
- Ministry of Finance Joint Development Project, Taipei
  财政主管部门土地合作开发项目
- 26 Training Building, China Grain Products R&D Institute, Taipei
  中华谷类食品工业技术研究所新建大楼
- CYCU Student Dormitory: Re Cheng Hall, Chuingli
  中原大学新中北学生宿舍：热诚楼

2018

- Fourth monograph titled *Cultivating Compassion*, is published
  第四本作品专书发行
- 27 TFC Commercial Office Building, Hsinchu
  台肥新竹 D7-A 商办大楼
- 28 Fubon Fuzhou Headquarters, Fujian, China
  富邦福州海西总部大楼
- Ally Logistics Yangmei Center
  永联杨梅物流中心
- 29 China Evangelical Seminary Bade Campus, Taoyuan
  中华福音神学研究学院

VIRTUAL ARCHITECTURE

2015

- Q Tainan Public Library
  台南市立图书馆总馆竞赛

2016

- R NTUST Dormitory, Taipei
  台湾科技大学学生宿舍
- Kang Chiao Bilingual School, Hsinchu
  新竹康桥校舍

Q

R

Project Credits

FORWARD

Sketch
Lang Chuang

Photographer
Bor-Nian Lin

LEGACY

Foundation

Photographer

Courtesy of Joshua Jih Pan and Chiu Hwa Wang

Formation

Photographer

Gina Lin, JJP; courtesy of NTOU, Joshua Jih Pan

Emergence

Photographer

Jeffrey Cheng, Yi-Wen Chen, Jim Chang, JJP

Culmination

Photographer

Rick Gardner, Yen-Chih Tseng, JJP; Courtesy of Taiwan Culture and Arts Foundation, Architizer, Yinqi Culture & Music Foundation, Delta Electronics Foundation

Review and Renew

Photographer

Te-Fan Wang, Bor-Nian Lin, Min-Hsiung Tseng, Jeffrey Cheng, JJP

VISION

[Discourse] The Inconstant and Constant in Architecture

Photographer

Rick Gardner, Weishih Hsieh, Yen-Chih Tseng, Ping Yang

Library of Public Information

Principal-In-Charge

Jason Chen

Project Team (Architectural/ Landscape/ Interior)

Joshua Jih Pan, Sheng-Tien Yeh, Shang-Ping Lin, Chi-Ming Chang, Chien-Yuan Liang, Vincent Wang, Chia-Jung Hsieh, Ady Tsai, Hsiao-Mei Shih, Yu-Ting Wu, Tien-Kai Yang, Shih-Fang Huang, Shih-Wei Liang, Han-Shen Chen, Chi-Hsuan Peng, Tien-Yi Pan, Kuan-Ju Hou, Genie Huang, Sara Hung, Chun-Lung Liu, Cong-Sian Chen, Sung-Po Huang, Jen-Chieh I, Chun-Sheng Lee, Wen-Sen Lin, Chu-Ching Chen, Chih-Jen Liu, Hui-Chi Chen, Chung-Ping Yeh, Cheng-En Wang, Cheng-Chih Kao, Shang-I Tsai

Client

Taichung Library

Consultant

Structural: Team Engineering Consulting Ltd.; MEP: Sine & Associates M/ELEC, Consultants & Engineers; Landscape: The Environmental Arts Design Company-master plan; Lighting: Chroma33 Architectural Lighting Design Inc.; Furniture: C. H. Wang, Architect; Architectural Façade System: Lead Dao Technology & Engineering Ltd.; Public Arts: Prof. Ming-Hong Yen

General Contractor

Kong Chou Construction Enterprise

Photographer

Weishih Hsieh, Kevin Wu

Chocolate Museum

Principal-In-Charge

Chungwei Su

Project Team (Architectural)

Joshua Jih Pan, Steven B.J. Chen, Gungo H.Y. Lin, Ady Tsai, Tien-Tso Huang, Vincent Wang, Hsiao-Mei Shih, Shih-Wei Liang, Den-Pu Yen

Client

Hunya Foods Co., Ltd.

Consultant

Structural: Team Engineering Consulting Ltd.; MEP: C. H. Wu Consulting Engineers; Landscape: Gardener Landscape & Construction Co., Ltd.; Interior: Horizon Design Co., Ltd.; Architectural Façade System: Lead Dao Technology & Engineering Ltd.

General Contractor

Sansin Builders Co., Ltd.

Photographer

Weishih Hsieh

Advantech Smart Campus

Principal-In-Charge

Jason Chen

Project Team (Architectural/ Landscape/ Interior)
Joshua Jih Pan, Tien-Kai Yang, Simon Y.T. Lin, Shun-Lang Wu, Hsin-Han Chen, Chia-Wei Ting, Po-Cheng Wang, Hsin-Ping Lee, Tien-I Pan, I-Chih Chen, Shu-Chiao Hsu, Wei-Lun Hsieh, Shih-Fang Huang, Chia-Ling Yang, Ching-Jen Chen, Chen-Yang Chang, Pao-Chao Hsu

Client

Advantech Co., Ltd.

Consultant

Phase I

Structural: Far Exp Engineering Consultants; MEP: Yuan Heng Electrical Engineers Office; Interior: Shaper Park Interior Design

Phase 2

Structural: Far Exp Engineering Consultants; MEP: Yuan Heng Electrical Engineers Office; Interior: Yuan Bang Interior Design

General Contractor

Phase1

Chung-Lin General Contractors, Ltd, Hao Chih Construction & Engineering Co., Ltd.

Phase 2

Chung-Lin General Contractors, Ltd., CTCI Smart Engineering Corporation, Town Chief Engineering Co., Ltd.

Photographer

Weishih Hsieh, Kando Chen, Jeffrey Cheng, Tien-Kai Yang

NCCU Research and Incubation Center

Principal-In-Charge

Joshua Jih Pan

Project Team (Architectural/ Landscape/ Interior)

Jason Chen, Shao-Ping Chen, Mitchell C.S. Huang, Yi-Shien Chin, Yuan-Pin Weng, Sheng-Ping Lin, Chun-Hua Liao, Pai-Wen Liao, Cheng-Hung Chen, Hung-Lin Wei, Che-Yi Liao, Yi-Li Li, Yu-Wei Lin

Client

Chengchi University

Consultant

Structural: Team Engineering Consulting Ltd.; MEP: C.H. Wu Consulting Engineers

General Contrator

Guo-Chi Construction Co., Ltd.

Photographer

Weishih Hsieh

Taiwan Photon Source Facility

Principal-In-Charge

Chung-Tsai Huang

Project Team(Architectural)

Chungwei Su, I-Li Mao, Steven B.J. Chen, Hsueh-Fen Chien, Shu-Chiao Hsu, Chia-Wei Ting, Sheng-Ping Lin, Hsiao-Yun Wu, Heng-Yuan Hung, Liang-Chen Hung, Chia-Chen Wang, Ching-Peng Chen, David Wei Gao, Mallet Y.L. Chang, Tsung-Yen Hsieh, Yi-Chyau Su, Fang-Cheng Wu, Cheng-En Wang, Chin-Chiu Chou, Cheng-Chih Kao, Wei-Chieh Liu, Hung-Lin Wei, Chih-Jung Chen, You-Jan Yan, Yen-Wei Chang, Shen-Che Su, Chao-I Liao, Chen-Yang Chang, Dan Xin

Client

Taiwan Synchrotron Radiation Research Center

Consultant

Structural: Moh and Associates, Inc.; MEP: Heng Kai Engineering Co., Ltd., Yuan Tai

Engineering Consulting Co., Ltd.; Micro Vibration: Tao Xu & Associates

General Contractor
Kedge Construction Co., Ltd., E&C Engineering Corporation

Photographer
Weishih Hsieh

Neo Solar Power Headquarters

Principal-In-Charge
Joshua Jih Pan

Project Team(Architectural/ Landscape)
Chungwei Su, Chung-Tsai Huang, Simon Y.T. Lin, Shun-Lang Wu, Martin Hagel, Mallet Y.L. Chang, Kun-Chao Lin, I-Fen Chang, Shu-Chiao Hsu, Chiu-Ping Wang, Chien-Hui Wu

Client
Neo Solar Power Corporation

Consultant
Structural: Team Engineering Consulting Ltd.; MEP: Home Run Electrical Engineers Associates

General Contractor
Fu Tsu Construction Co., Ltd., Lead Fu Industrials Corp.(curtain wall)

Photographer
Jeffrey Cheng

Ally Logistics Yangmei Center

Principal-In-Charge
Chung-Tsai Huang

Project Team(Architectural/ Landscape)
Joshua Jih Pan, Mitchell C.S. Huang, Hsiao-Ming Chang, Chao-Yi Chen, Ady Tsai, Po-Yang Lin, Teng-Mao Huang, Pei-Shan Chang, Min Chiu, Hsin-Han Chen, Po-Cheng Wang, Chun Ouyang, Syue-Shuang Leng, Ming-Yu Wei, Shih-Fang Huang, Huei-Ru Tai, Chien-Lin Chen

Client
Cathay Life Insurance, Ally Logistic Property Co., Ltd.

Consultant
Structural: Envision Engineering Consultants; MEP: Cozeta Energy Service Corp.; Architectural Facade System: Chungan Wellsun Co., Ltd.

General Contractor
Xu Yuan Construction Co., Ltd.

Photographer
Tzuu-Lieh Tsai, Ping Yang

Giant Bicycles Headquarters

Principal-In-Charge
Joshua Jih Pan

Project Team (Architectural/ Landscape/ Interior)
David Lee, Yang-Sheng Chen, Ady Tsai, Chao-Ching Wang, Syue-Shuang Leng, Hsin-Ping Lee, Shih-Fang Huang, Kuo-Hao Hsu, Cheng-Chih Kao, Shao-Hsiang Wei, Tzuu-Lieh Tsai, Shih-Chi Sung

Client
Giant Taiwan Co., Ltd.

Consultant
Structural: Envision Engineering Consultants; MEP: C.H. Wu Consulting Engineers; Lighting: Amphlilight Design Studio; Green: Greentend Constulant; Geo-tech: Chung Chi Technical Consultant Co., Ltd.

General Contractor
Lee Ming Construction Co., Ltd

NCCU Center for Public and Business Administration Education

Principal-In-Charge
Chungwei Su

Project Team (Architectural/ Landscape/ Interior)
Joshua Jih Pan, Sheng-Tien Yeh, Vincent Wang, Rafael Martinez, Yu-Ju Cheng, Wei-Che Tang, Yuan-Pin Weng, Nan-Tse Su, Frank C.C. Liao, Ming-Yu Wei, Shih-Fang Huang, Chia-Ling Yang, Ke-Hang Shen

Site Supervision / PCM
Yu, Tai-Sheng Architects & Associates

Client
Chengchi University

Consultant
Structural: Far Exp Engineering Consultants; MEP: Jehng Long Engineering Co., Ltd.; Acoustic: In Li Chia Co., Ltd.

General Contractor
Li Jin Engineering Co., Ltd.

SOCIAL IMPACT

[Dialogue] Cui Kai x Joshua Jih Pan: Cross-strait Architectural Practice Environment and Social Impacts

Photographer
Odie Yang, Tingda Liang

Taoyuan International Airport MRT Stations

Design Architect (A1 Station)
Maki and Associates, Japan

Principal-In-Charge
Chungwei Su, Tse-Hsiung Cheng

Project Team (Architectural/ Landscape/ Interior)
A1 Station
Joshua Jih Pan, Jui-Min Huang, Vincent Wang, Hsiao-Ming Chang, Tien-Kai Yang, Ady Tsai, Mitchell C.S. Huang, Shih-Fang Huang, Chih-Ming Chang, Ping-Jung Ke, Chung-Yung Lin, Yu-Wei Mao, Wei-Hsiang Sung, I-Yu Chung, Chi-Hsiang Chen, Chun-I Lee

A2 to A6 Stations
Chia-Lin Liu, Sung-Ju Chen, Chih-Yun Chiang, Chih-Ming Chang, Siang-Yu Liang, Ady Tsai, Fu-Chih Tang, Tzu-Hua Wu, Ray-Heng Lu, Cheng-Kan Lin, Chia-Wei Ting, Chien-Yu Chen, David B.D. Wu, Wei-Ke Kuo, Shih-Fang Huang, Hui-Ru Tai, Chien-Yuan Liang, Shun-Yi Cheng, San-Min Lee, Bo-Yuan Hsiung, Su-Chuan Wang

A22, A23 Stations
Chih-Ming Chang, Siang-Yu Liang, Chih-Yun Chiang, Tien-I Pan, Yi-Shien Chin, Chia-Jung Hsieh, Chia-Cheng Wei, Tien-Tso Huang, Chao-Yen Kang, Jui-Min Huang, Mitchell C.S. Huang, Chia-Ling Yang

Client
A1Station: Department of Rapid Transit Systems, Taipei City Government
A2 to A6, A22, A23 Stations: Bureau of High Speed Rail, Ministry of Transportation and Communications

Consultant
A1 Station
Structural: CECI Engineering Consultants, Inc., Taiwan, Evergreen Consulting Engineering, Inc., Tomita Structural Design, Envision Engineering Consultants; MEP: CECI Engineering Consultants, Inc., Taiwan; Lighting: Chroma33 Architectural Lighting Design Inc.; Architectural Façade System: Origin Design & Engineering Co., Ltd.; Geo-tech: CECI Engineering Consultants, Inc., Taiwan

A2, A3, A4 Stations
Structural: Team Engineering Consulting Ltd. MEP: CECI Engineering Consultants, Inc., Taiwan

A5, A6 Stations
Structural: CECI Engineering Consultants, Inc., Taiwan; MEP: CECI Engineering Consultants, Inc., Taiwan

A22 Station
Structural: Envision Engineering Consultants; MEP: CECI Engineering Consultants, Inc., Taiwan

A23 Station
Structural: Team Engineering Consulting Ltd.; MEP: CECI Engineering Consultants, Inc., Taiwan

General Contractor
A1: Fu Tsu Construction Co., Ltd.
A2, A3, A4: Da Cin Construction Co., Ltd.
A5, A6: Hwang Chang General Contractor Co., Ltd.
A22, A23: Continental Engineering Corporation

Photographer
Weishih Hsieh, Chao-Chih Chou

When the Bells Toll: Chang Yuan Christian University Series

Chang Ching Yu Memorial Library

Principal-In-Charge
Joshua Jih Pan

Joint Venture Architect
Chih Hwa Wang

Project Team (Architectural/ Landscape/ Interior)
Steve Kao Hsu, Jason Chen, Chun-Fa Tan, Tsai-Mao Wang, Tse-Hsiung Cheng, Jong-Yu Cheng, Lang Chuang, Cheng-Fan Liu, Yao-Hsin Tang, Tatung Chou

Client
Chung Yuan Christian University

Consultant
Structural: Wen-Liang Hsu Architects; MEP: Zhu Yu Engineering Consulting Co., Ltd., I. S. Lin & Associates

General Contractor
Jing Cheng Construction Co., Ltd., Yu Tian Construction Co., Ltd.

Photographer
Freddy Chung

Holistic Education Village

Principal-In-Charge
Joshua Jih Pan

Joint Venture Architect
Chiu Hwa Wang

Project Team (Architectural/ Landscape/ Interior)
Steve Kao Hsu, Johnson C.P. Chiang, Ady Tsai, Robin C. K. Tang, Simon Y. T. Lin, Steven B.J. Chen, Don K. Marchildon, Chien-Ming Hsu, Yun-Ju Chien, Pei-Shih Chao, Chia-Lin Hsu, Tzuu-Lieh Tsai, I-Pin Kuo, Hung-Hsun Li, Jen-Hui Wu

Client
Chung Yuan Christian University

Consultant
Structural: Horngyun Structural Engineering Consultant; MEP: Elite Consultant Engineers & Associates, Yuan Tai Engineering Consulting Co., Ltd. ; Lighting: Chroma33 Architectural Lighting Design Inc.; Acoustic: Prof. Wei-Hwa Chiang

General Contractor
Sun Sea Construction Co., Ltd., Chung Kuang Construction Co., Ltd.

Photographer
Jeffrey Chang

Student Dormitory: Re Cheng Hall

Principal-In-Charge
Joshua Jih Pan

Project Team (Architectural/Landscape/ Interior)
Chungwei Su, Chung-Tsai Huang, Yen-Chih Tseng, Steven B. J. Chen, Liang-Tzu Chen, Ling Lin, Wei-Lun Chen, Ying-Ting Huang, Shih-Fang Huang, Chia-Ling Yang, Fang-Cheng Wu, Yen-Wei Chang, Kuo-Long Tsao, Hsiao-Wen Lee, Pao-Chao Hsu, Te-Chia Lee

Client
Chung Yuan Christian University

Consultant
Structural: Team Engineering Consulting Ltd.; MEP: C.H. Wu Consulting Engineers, Yuan Tai Engineering Consulting Co., Ltd.; Lighting: GIIDPLUS Design Studio

General Contractor
Sun-Sea Construction Corporation

Photographer
Weishih Hsieh, Kevin Wu, Yen-Chih Tseng

Mackay Medical College

Principal-In-Charge
Joshua Jih Pan

Project Team (Architectural/Landscape/ Interior)
Chung-Tsai Huang, Shao-Ping Chen, Yen-Chih Tseng, Yuan-Pin Weng, Yi-Shien Chin, Wei Su, Su-Chuan Wang, Shih-Fang Huang, Chia-Ling Yang, Mitchell C.S. Huang, Po-Yang Lin, Tzuu-Lieh Tsai, Yi-Sung Liu, Yen-Chang Chen

Client
Mackay Medical College

Consultant
Structural: Envision Engineering Consultants; MEP: C.H. Wu Consulting Engineers, Yuan Tai Engineering Consulting Co., Ltd.

General Contractor
Li Jin Engineering Co., Ltd.

Photographer
Weishih Hsieh, Yen-Chih Tseng

Xinglong Public Housing Block 1

Principal-In-Charge
Jason Chen

Project Team (Architectural/Landscape/ Interior)
Joshua Jih Pan, Steven B.J. Chen, Chih-Ming Chang, Mallet Y.L. Chang, Yang-Sheng Chen, Lin Ling, Tzu Yuan Tsai, Po-Chang Wu, Shih-Fang Huang

Site Supervision
Yu, Tai-Sheng Architects & Associates

Client
Department of Urban Development, Taipei City Government

Turnkey Contractor
Ruentex Corp.

Photographer
Chien-Yuan Liang, Yang-Sheng Chen, Mallet Y.L. Chang

Ministry of Finance Joint Development Project

Principal-In-Charge
Jong-Yu Cheng

Project Team (Architectural/Landscape/ Interior-office)
Joshua Jih Pan, Hsiao-Ming Chang, Oui Ming Sae Tang, Chien-Yuan Liang, Hsueh-Fen Chien, Su-Chuan Wang, Li-Jiun Lu, Mien-Li Hsieh, Ying-Jie Chen, Shih-Fang Huang, An-Yuan Chung, Chia-Ling Yang, Yen-Lin Huang, Hsiao-Mei Shih, Chun-Ting Chang, Chia-Wei Ting, Tien-Yi Pan, Tien-Kai Yang, Chi-Cheng Chuang, Ching-Feng Shen, Hung-Lin Wei, Shao-Hsiang Wei, Cheng-Hung Chen, Yen-Chang Chen, Tzu-Ying Yang, Chia-Chen Wang

Client
Ministry of Finance; Huaku Development Co., Ltd.

Consultant
Structural: Team Engineering Consulting Ltd.; MEP: C.H. Wu Consulting Engineers(office), Sine & Associates M/ELEC Consultants & Engineers(housing); Lighting: GIIDPLUS Design Studio; Interior: JAHAA Design(housing)

General Contractor
Pin-Sing Construction Co., Ltd. (office), Continental Engineering Corporation(housing)

Photographer
Hsiao Hsiung Liang Yen, Johnny Yang, Chien-Yuan Liang

Bethany Children's Home

Principal-In-Charge
Joshua Jih Pan

Project Team (Architectural/Landscape/ Interior)
Jong-Yu Cheng, Shao-Ping Chen, Yuan-Pin Weng, Shih-Fang Huang, Yi-Hsueh Lai, Po-Lung Chang, Yen-Chang Chen

Client
Bethany Children's Home

Consultant
Structural: Team Engineering Consulting Ltd.; MEP: C.H. Wu Consulting Engineers

General Contractor
Sansin Builders Co., Ltd.

Campus Evangelical Fellowship Building

Principal-In-Charge
Joshua Jih Pan

Project Team (Architectural/Landscape/ Interior)
Tien-Kai Yang, Rafael Martinez, Ming-Yu Wei, Shih-Fang Huang

Client
Campus Evangelical Fellowship of Taiwan

Consultant
Structural: Envision Engineering Consultants; MEP: C.H. Wu Consulting Engineers

General Contractor
Li Jin Engineering Co., Ltd.

Center for Advanced Cancer Therapies

Principal-In-Charge
Chungwei Su

Project Team (Architectural/ Landscape/ Interior)
Cancer Center
Joshua Jih Pan, Hsiao-Ming Chang, Oui Min Sae Tang, Wen-Chih Wang, Hsueh-Fen Chien, Chien-Yuan Liang, Ying-Jie Chen, Mien-Li Hsieh, Shih-Fang Huang, Su-Ching Wu, Yen-Chih Tseng, Kevin F. Hsieh, Hsuan-Yang Mao, Yen-Hung Liu, Chih-Hsuan Peng, I-Hao Chen, Yu-Wei Mao, Lien-Sheng Chen, Fang-Cheng Wu, Chiu-Ping Wang, Tzuu-Lieh Tsai, Den-Pu Yen

Radiation and Proton Therapy Center
Oui Ming SaeTang, Yen-Lin Huang, Hsiao-Mei Shih, Po-Sung Yen, Wei-Che Tang, Huei-Ru Tai, Nan-Hung Chen, Den-Pu Yen

Client
YongLin Healthcare Foundation

Medical Planning/Schematic Designer
Cancer Center: HKS Architects, US

Consultant
Cancer Center
Structural: Team Engineering Consulting Ltd.; MEP: Hander Engineering & Construction Inc., Yuan Tai Engineering Consulting Co., Ltd.; Landscape: Environmental Arts Design Company; Lighting: GIIDPLUS Design Studio; Interior: JAHAA Design

Radiation and Proton Therapy Center
Structural: Team Engineering Consulting Ltd.; MEP: Minginer Consultant Co., Ltd., Yuan Tai Engineering Consulting Co., Ltd.; Radiation: Feynman Technology Co., Ltd.; Proton Specialist: Bard, Rao + Athanas Consulting Engineers, P.C.; BIN: WeBin Services Co., Ltd.; Flood Protection: Yuan Feng Engineering Co., Ltd.

General Contractor
Cancer Center
Da Cin Construction Co., Ltd.

Radiation and Proton Therapy Center
Shimizu Corporation, Taipei Office

SUSTAINABILITY

[Dialogue] Wim Chang x Chungwei Su: The Perfect Architect-Client Collaboration

Photographer

Wei-Kang Wang, Delta Electronics Foundation, Jay Liou, Guan-Liang Liou, JJP

Architectural Terroir: Delta Electronics Series

Ring of Celestial Bliss, Sports Court Shelter of Daxi Elementary School

Principal-In-Charge

Joshua Jih Pan

Project Team

Chungwei Su, Steven B.J. Chen, Yen-Chih Tseng, Yen-Wei Chang, Guan-Liang Liou(shelter)

Client

Delta Electronics Foundation

Consultant

Structural: Envision Engineering Consultants

General Contractor

Li Jin Engineering Co., Ltd., Sin Lun Construction Co., Ltd.(shelter), Jih Jian Industrial Co., Ltd.(shelter)

Photographer

Delta Electronics Foundation, Yen-Chih Tseng, Guan-Liang Liou

Delta Tainan Plant Phase II

Principal-In-Charge

Chungwei Su

Project Team (Architectural)

Joshua Jih Pan, Steve Kao Hsu, Liang-Chen Hung, Wu-Chih Chen, Chia-Hung Tsai

Client

Delta Electronics, Inc., Taiwan

Consultant

Structural: TEAM Engineering Consulting Ltd.; MEP: Hsueh-Shun Wang (Cozeta Energy Service Corp.); Landscape: Manor Landscape Engineering Co., Ltd

General Contractor

Li Jin Engineering Co., Ltd.

Photographer

Jay Liou, Delta Electronics, Inc., Liang-Chen Hung

Delta Americas Headquarters

Principal-In-Charge

Joshua Jih Pan

Project Team (Architectural/ Landscape)

Steve Kao Hsu, David Lee, Shih-Fang Huang, Tsung-Shun Yeh, Po-Chang Wu, Yang-Sheng Chen

Client

Delta Electronics (Americas) Ltd.

Architect of Record

Korth Sunseri Hagey Architects

Consultant

Structural: Nishkian Menninger; MEP: Timmons Design Engineers; Landscape: SWA Group

General Contractor

Vance Brown Builders

Photographer

Rick Gardner, David Lee

Taichung First High School Heritage Building

Principal-In-Charge

Chungwei Su

Project Team (Architectural/ Landscape)

Joshua Jih Pan, Steven B.J. Chen, Chao-Yen Kang, Chao-Yi Chen, Chia-Cheng Wei, Wei-Chieh Liu

Client

Delta Electronics Foundation

Consultant

Historical Preservation: Ya-Ping Lin Architect of Wilderness Field; Structural: TEAM Engineering Consulting Ltd.; MEP: Sound Engineering Consulting Ltd.; Lighting: Amphlilight Design Studio

General Contractor

Tai-Nan Construction Co., Ltd.

Photographer

Jay Liou, Taichung First High School

Horizon Design Shanghai Office

Principal-In-Charge

Joshua Jih Pan

Project Team (Architectural/ Landscape/ Interior)

Yu-Ting Wu, Jong-Yu Cheng, Rocky Huang, Jen-Chieh Cheng, Ady Tsai, Jason C.L. Wu, Li-Kai Lee, Yu-Lin Lee

Client

Horizon Design Co., Ltd.(Shanghai)

Consultant

Structural: Team Engineering Consulting Ltd., Shanghai Long Heng Decoration Co., Ltd.; MEP: Shanghai JinRi M&E Installing Co., Ltd

General Contractor

Shanghai Long Heng Decoration Co., Ltd.

Photographer

Jim Chang, Steve Mok, Heng-Zhong Lu, JJP

---

tsmc Campus

Principal-In-Charge

Fab12, Fab15: Joshua Jih Pan
Fab14: Jong-Yu Cheng

Project Team (Architectural/ Landscape)

Fab12 Headquarters

Oui Ming Sae Tang, Tatung Chou, Wen-Chung Tseng, Kevin F. Hsieh, Shih-Ming Huang, Henry W.Y. Yen, Chi-Te Sung, Shu-Hui Lin, Chih-Wei Tsai, Kuan-Ju Hou, Yung-Bin Huang, Ming-Cheng Lai, Ching-Feng Shen, Shou-Ming Hsieh, Cheng-Wei Li

Fab12 Phase 4

Chih-Wei Tsai, Steven B.J. Chen, Liang-Chen Hung, Chien-Hung Shen, Shun-Lang Wu, Kun-Chao Lin, Shu-Chiao Hsu, Kuan-Ju Hou, Chun-Kai Shih, Shih-Fang Huang, Yung-Bin Huang, Yi-Chyau Su, Hung-Ta Hung, Wei-Fu Shih, Shun-I Yu, Yi-Sung Liu, Cheng-Hung Su, Kuang-Wu Hou, Kuo-Hao Hsu, Meng-Chieh Chen, Yung-Ching Pan

Fab14 Phase 5

Joshua Jih Pan, Henry W.Y. Yen, Chih-Wei Tsai, Yen-Chih Tseng, Kuan-Ju Hou, Shun-Yi Cheng, I-Li Mao, Shu-Ching Wu, Tsung-Yin Lee, Li-Jiun Lu, I-Fen Chang, Siang-Yu Liang, Ming-Hsien Yu, Chao-Yen Kang, Yu-Hsuan Chu, Shih-Fang Huang, Yu-Lin Lee, Ching-Feng Shen, Cong-Sian Chen, Sung-Po Huang, Chia-Hung Tsai, Guo-Wei Lu, You-Jan Yan, Jr-Cheng Syu, Chia-Ming Yang, Pei-Chao Hsu, Pao-Chao Hsu, Shao-Hsiang Wei, Teng-Hung Wang, Chia-Hao Li, Meng-Min Kuo

Fab15 Phase 1

Oui Ming Sae Tang, Pen Lee, Yen-Chih Tseng, Hsuan-Ming Wu, Kevin F. Hsieh, Hsiao-Yun Wu, Chun-Chih Liu, Wen-Chih Wang, Yen-Lin Huang, Judy Yang, Frank C.C. Liao, Mien-Li Hsieh, Henry W.Y. Yen, Shih-Fang Huang, Tsung-Yin Lee, Li-Jiun Lu, Tzu-Hua Wu, Siang-Yu Liang, Yang-Sheng Chen, Chi-Hsuan Peng, Chien-Hung Shen, Yen-Hung Liu, Po-Yang Lin, Shih-Wei Liang, Chi Hsiao, Wen-Yu Liao, Ang Lu, Cheng-Ying Tsai, Pao-Chao Hsu, Yung-Bin Huang, Hung-Ta Hung, Che-Yi Liao, Jr-Cheng Syu, Chun-Lung Liu, Wei-Chieh Liu, Sung-Po Huang, You-Jan Yan, Meng-Min Kuo, Ting-Han Lin, Shao-Hsiang Wei, Chia-Ming Yang, Shyun-Ming Huang, Shih-Min Lee, Kuang-Chi Huang

Client

Taiwan Semiconductor Manufacturing Company

Consultant

Fab12 Headquarters

General Consultant: IDC Taiwan Inc.; Structural: Goodtech Engineering Consultant Company; MEP: Angel Engineering Consultants; Landscape: Environment Arts Design Company; Lighting: Chroma33 Architectural Lighting Design Inc.; Water & Soil Conservation: Chung Chi Technical Consultant Co., Ltd., Interior Design: Mandartech Interiors Inc.

Fab12 Phase 4

Structural: Team Engineering Consulting Ltd.; Water & Soil Conservation: Chung Chi Technical Consultant Co., Ltd.

Fab14 Phase 5

Structural: Team Engineering Consulting Ltd.; MEP: E&C Engineering Corporation(office), United Integrated Services Co., Ltd.(fab); Landscape: Old Farmer Landscape Architecture Co., Ltd., Geo-Tech Conservation: Chung Chi Technical Consultant Co., Ltd.; Interior Design: Mandartech Interiors Inc. (office)

Fab15 Phase 1

Structural: Team Engineering Consulting Ltd.; MEP: Acter Co., Ltd.(office), United Integrated Services Co., Ltd.(fab); Landscape: Old Farmer Landscape Architecture Co., Ltd.; Water & Soil Conservation: Chung Chi Technical Consultant Co., Ltd.

General Contractor

Fab12 Headquarters

Fu Tsu Construction Co., Ltd., Lead Fu Industrial Co., Ltd.(curtain wall), China Steel Structure Co., Ltd.(steel structure)

Fab12 Phase 4

Fu Tsu Construction Co., Ltd.; Tasa Construction Corp.(office), Da Cin Construction Co., Ltd.(pipe bridge)

Fab14 Phase 5

Tasa Construction Corp.(office), Lead Fu Industrial Co., Ltd.(office wall), Fu Tsu Construction Co., Ltd.(fab), Da Chin Construction Co., Ltd.(fab), I-Domain Industrial Co., Ltd.(fab wall), China Steel Structure Co., Ltd. (steel structure)

Fab15 Phase 1

Da Cin Construction Co., Ltd.(office); Mega Façade Corp.(office wall); Fu Tsu Construction

Co., Ltd. (fab),
I-Domain Industrial Co., Ltd. (fab wall), Taiwan Sto Architectural Co., Ltd.(fab wall), China Steel Structure Co., Ltd.(steel structure)

Photographer

David Chen, Jeffrey Cheng, Jay Liou

Aging Gracefully: Suang-Lien Series and Interdisciplinary Workshop

Suang-Lien Senior Living Community

Principal-In-Charge

Phase1: Joshua Jih Pan;
Phase2, 3: Chung-Tsai Huang

Project Team (Architectural/ Phase2-3 Landscape/ Phase2-3 Interior)

Phase1: Robin C.K. Tang, Hsueh-Fen Chien, Kevin F. Hsieh, Hui-Yu Kuan, Chih-Ming Lee, Tzuu-Lieh Tsai

Phase2: Robin C.K. Tang, Hsiao-Mei Shih, Ya-Jen Chen, Han-I Chu, Tzuu-Lieh Tsai

Phase3: Chungwei Su, Steven B.J. Chen, Weishih Hsieh, Pen Lee, Hsiao-Yuen Wu, Liang-Chen Hung, Hsuan-Ming Wu, Chien-Shen Liu, Yi-Sung Liu, Kuo-Long Tsao, Tzuu-Lieh Tsai

Client

Taiwan Presbyterian Suang-Lien Church

Consultant

Phase1
Structural: Goodtech Engineering Consultant Company; MEP: Teddy Associates Consultant Ltd., I. S. Lin and Associates; Landscape: Chiang Yen Environment Consulting Ltd.; Interior: Horizon Design Co., Ltd.

Phase 2
Structural: Team Engineering Consulting Ltd.; MEP: Elite Consultant Engineer & Associates, I. S. Lin and Associates,

Phase 3
Structural: Team Engineering Consulting Ltd; MEP: C.H. Wu Consulting Engineers, I. S. Lin and Associates; Lighting: GIIDPLUS Design Studio(church); Architectural Façade System: Lead Dao Technology & Engineering Ltd.(church); Acoustic: Prof. Wei-Hwa Chiang(church)

General Contractor

Phase1, 2: Sunbuild Construction Co., Ltd
Phase3: Gin Zon Construction Co., Ltd., Yuan Den Engineering Co., Ltd.

Photographer

Min-Hsiung Tseng, David Chen, Yi-Wen Chen, Weishih Hsieh, Steven B.J. Chen

Suang-Lien Xinzhuang Social Welfare Center

Principal-In-Charge

Chung-Tsai Huang

Project Team (Architectural/ Landscape)

Chungwei Su, Steven B.J. Chen, Sheng-Tien Yeh, Tien-Tso Huang, Guan-Liang Liou, Chia-Ling Yang, Shu-Han Hsu, Yi-Sung Liu

Client

Suang-Lien Social Welfare Foundation

Consultant

Structural: Team Engineering Consulting Ltd.; MEP: C.H. Wu Consulting Engineers; Interior: Weishih Hsieh Architect

General Contractor

UOBO Construction Co., Ltd.

Photographer

Weishih Hsieh

Interdisciplinary Workshop: Next Generation of Senior Living

Principal-In-Charge

Chung-Tsai Huang

Project Member

Joshua Jih Pan, Chungwei Su, Genie Huang, Ruyun Feng, Tingda Liang, Yang-Sheng Chen, Frank C.C. Liao, Steven B.J. Chen, Chi-Yu Yu, Pei-Shan Chang, Syue-Shuang Leng, Yen-Chih Tseng, Jen-Chieh Cheng, Tien-Kai Yang, Mitchell C.S. Huang, Youhan Hu, Zhe-Hong Tan

Developer

CMP Group Inc., PUJEN Land Development, CMP Pujen Foundation for Arts and Culture

Architect

Weishih Hsieh Architects, CCL Architects & Planners

Scenario Planning Facilitator

Scenario Lab (DJ Yu, Grace Lin, Céline Chen), Foundation of Universal Design Education (Feng-Chen Tang, Dedo Wang)

Professor and Expert

Prof. Shu-Chang Kung, Prof. Ming-Da Lee, Suang-Lien Center for the Elderly (Fan-Wen Tsai, Meng-Wei Tsai),
Jack C.C. Chen

Venue

The Jen Library; PUJEN Land Development; Shangri-La Paradise, Taiwan; CMP Block Museum of Arts;Taiwan Air Force Innovation Base

Photographer

Yen-Chih Tseng, Yang-Sheng Chen

SHARING

An Organic Office

Photographer

Jeffrey Cheng, Kevin Wu, Min-Hsiung Tseng, Weishih Hsieh, Wei-Kang Wang, Daniel Y. Tsai, Chien-Yuan Liang, Yen-Chieh Tseng, Shyue-Shuang Leng, JJP

Sketch

Ping Chen

Biographies

Photographer

Te-Fan Wang

Timeline

Photographer

Gina Lin, Min-Hsiung Tseng, Jeffrey Cheng, David Chen, Jay Liou, Weishih Hsieh, Yi-Wen Chen, Webber Huang, Hsiao-Ming Chang, Yen-Chih Tseng, JJP

项目信息

前言

手绘
庄朖

摄影师
林栢年

传承

建筑启蒙

影像提供
潘冀、王秋华

自立门户

摄影师
林淑卿、JJP

影像提供
台湾海洋大学、潘冀

在地深耕

摄影师
郑锦铭、陈逸文、张全箴、JJP

续攀高峰

摄影师
Rick Gardner、曾彦智、JJP

影像提供
台湾文化艺术基会、Architizer、音契文化音乐基金会、台达电子文教基金会

继往开来

摄影师
汪德范、林栢年、曾敏雄、郑锦铭、JJP

视角

建筑的变与不变

摄影师
Rick Gardner、谢伟士、曾彦智、杨平

公共资讯图书馆

项目主持人
陈浩生

项目成员（建筑 / 景观 / 室内）
潘冀、叶胜钿、林圣平、张志明、梁建元、王文錤、谢佳融、蔡恒升、史晓梅、吴昱廷、杨天闿、黄诗芳、梁世伟、陈瀚申、彭琦瑄、潘天壹、侯冠如、黄巾倪、洪怡姗、刘俊龙、陈聪贤、黄松柏、易人杰、李春生、林文森、陈取静、刘志仁、陈惠琪、叶中平、王承恩、高成志、蔡尚益

业主
台中图书馆

顾问
结构：群策工程顾问股份有限公司；机电：正弦工程顾问有限公司；景观：环艺工程顾问有限公司（前期规划）；灯光：大公照明设计顾问有限公司；家具：王秋华建筑师；外墙分析：利道科技工程有限公司；广义公共艺术：颜名宏教授

施工方
港洲营造股份有限公司

摄影师
谢伟士、吴启民

巧克力博物馆

项目主持人
苏重威

项目成员（建筑）
潘冀、陈伯桢、林泓宇、蔡恒升、黄天佐、王文錤、史晓梅、梁世伟、严丹璞

业主
宏亚食品股份有限公司

顾问
结构：群策工程顾问股份有限公司；机电：吴建兴机电空调技师事务所；景观：大乔木景观工程有限公司；室内：汉象设计工程股份公司；外墙分析：利道科技工程有限公司

施工方
三星营造股份有限公司

摄影师
谢伟士

研华智能园区

项目主持人
陈浩生

项目成员（建筑 / 景观 / 室内）
潘冀、杨天闿、林镱汰、吴顺郎、陈鑫翰、丁家伟、汪栢成、李欣屏、潘天壹、陈镱铨、许舒乔、谢维伦、黄诗芳、杨家琳、陈景仁、张宸样、许堡朝

业主
研华股份有限公司

顾问
一期：结构：远拓顾问有限公司；机电：元亨电机技师事务所；室内：乔杰室内装修设计公司
二期：结构：远拓顾问有限公司；机电：元亨电机技师事务所；元邦室内装修设计公司

施工方
一期：中麟营造股份有限公司、豪志工程股份有限公司
二期：中麟营造股份有限公司、益鼎工程股份有限公司、东正工程股份有限公司

摄影师
谢伟士、陈耀钦、郑锡铭、杨天闿

政治大学研究暨创新育成总中心

项目主持人
潘冀

项目成员（建筑 / 景观 / 室内）
陈浩生、陈绍平、黄介帅、覃亦仙、翁源彬、苏玮、黄诗芳、林圣平、廖俊华、廖珮雯、陈政宏、魏鸿林、廖哲逸、李怡丽、林育纬

业主
政治大学

顾问
结构：群策工程顾问股份有限公司；机电：吴建兴电机空调技师事务所

施工方
国记营造股份有限公司

摄影师
谢伟士

台湾光子源加速器实验中心

项目主持人
黄种财

项目成员（建筑）
苏重威、毛以励、陈伯桢、简雪芬、许舒乔、丁家伟、林圣平、吴晓云、洪珩元、洪亮辰、王家珍、陈敬鹏、高卫、张元琳、谢宗晏、苏怡樵、吴方诚、王承恩、邱钦洲、高成志、刘为杰、魏鸿林、陈志荣、颜佑展、张晏维、苏圣喆、廖超逸、张宸样、辛丹

业主
台湾同步辐射研究中心

顾问
结构：亚新工程顾问股份有限公司；机电顾问：恒开工程顾问有限公司；空调：元泰工程顾问有限公司；微振：Tao Xu & Associates

施工方
根基营造股份有限公司、益鼎工程股份有限公司

摄影师
谢伟士

新日光能源总部大楼

项目主持人
潘冀

项目成员（建筑 / 景观）
苏重威、黄种财、林镱汰、吴顺郎、贺马汀、张元琳、林崑炤、张懿芬、许舒乔、黄诗芳、王秋斌、吴建辉

业主
新日光能源科技股份公司

顾问
结构：群策工程顾问股份有限公司；机电：宏伦电机工业技师事务所

施工方
互助营造股份有限公司、力福实业股份有限公司（帷幕墙）

摄影师
郑锦铭

永联杨梅物流中心

项目主持人
黄种财

参与成员
潘冀、黄介帅、张晓鸣、陈昭邑、蔡恒升、林柏阳、黄登懋、张佩珊、邱敏、陈鑫翰、汪柏成、欧阳钧、冷雪霜、魏名妤、黄诗芳、戴慧如、陈建霖

业主
国泰人寿保险股份有限公司、永联物流开发股份有限公司

顾问
结构：筑远工程顾问有限公司；机电：高齐能源科技股份有限公司；外墙分析：强安威胜股份有限公司

施工方
旭源营造工程股份有限公司

摄影师
蔡祖烈

巨大机械总部大楼

项目主持人
潘冀

项目成员（建筑 / 景观 / 室内）
李大威、陈扬升、蔡恒升、王朝庆、冷雪霜、李欣屏、黄诗芳、徐国浩、高成志、魏绍翔、蔡祖烈、宋士奇

业主
巨大机械工业股份有限公司

顾问
结构：筑远工程顾问有限公司；机电空调：吴建兴机电空调技师事务所；灯光：颉乘设计照明工作室；绿建筑：岩仲绿建筑顾问有限公司；大地：中基土壤技术顾问有限公司

施工方
丽明营造股份有限公司

政治大学公共行政及企业管理教育中心

项目主持人
苏重威

项目成员（建筑 / 景观 / 室内）
潘冀、叶胜钿、王文錤、马法耶、郑汝如、唐伟哲、翁源彬、苏楠泽、廖志坚、魏名妤、黄诗芳、杨家琳、沈可航

监造 / 专案管理
喻台生建筑师事务所

业主
政治大学

顾问
结构：远拓顾问有限公司；机电空调：正龙工程股份有限公司；声学：音利佳股份有限公司

施工方
利晋工程股份有限公司

社会影响

崔恺 x 潘冀：两岸建筑执业环境差异与社会影响

摄影师
杨松华、梁廷达

台湾桃园国际机场捷运车站

设计建筑师（A1 站）
日本桢综合计划事务所

项目主持人
苏重威、郑泽雄

项目成员（建筑 / 景观 / 室内）
A1 站
潘冀、黄瑞民、王文錤、张晓鸣、杨天闿、蔡恒升、黄介帅、黄诗芳、张志明、柯炳荣、林仲雍、毛语薇、宋伟祥、钟易佑、陈启祥、李俊仪

A2–A6 站
刘家麟、陈松如、姜智匀、张志明、梁向宇、蔡恒升、汤富智、吴姿桦、卢瑞恒、林正侃、丁家伟、陈建宇、吴柏德、郭玮克、黄诗芳、戴慧如、梁建元、郑顺益、李三民、熊博远、王素娟

A22–A23 站
张志明、梁向宇、姜智匀、潘天壹、覃亦仙、谢佳融、魏嘉成、黄天佐、康兆言、黄瑞民、黄介帅、杨家琳

业主
A1 站：台北市政府捷运工程局
A2 - A6, A22, A23 站：交通部铁道局

顾问
A1 站
结构：台湾世曦工程顾问股份有限公司、永峻工程顾问股份有限公司、富田构造设计事务所、筑远工程顾问有限公司；机电：台湾世曦工程顾问股份有限公司；灯光：大公照明设计顾问有限公司；外墙分析：佑捷设计工程有限公司；大地：台湾世曦工程顾问股份有限公司

A2, 3, 4 站
结构：群策工程顾问股份有限公；机电：台湾世曦工程顾问股份有限公司；

A5, A6 站
结构：台湾世曦工程顾问股份有限公司；机电：台湾世曦工程顾问股份有限公司

A22 站
结构：筑远工程顾问有限公司；机电：台湾世曦工程顾问股份有限公司

A23 站
结构：群策工程顾问股份有限公；机电：台湾世曦工程顾问股份有限公司

施工方
A1：互助营造股份有限公司
A2-A3-A4：达欣工程股份有限公司
A5-A6：皇昌营造股份有限公司
A22-A23：大陆工程股份有限公司

摄影师
谢伟士、周兆志

当钟声响起：中原大学校园建筑系列

张静愚纪念图书馆

项目主持人
潘冀

合作建筑师
王秋华

项目成员（建筑 / 景观 / 室内）
许国、陈洁生、谭俊发、王才茂、郑泽雄、郑荣裕、庄脤、刘正帆、汤耀欣、周大同

业主
中原大学

顾问
结构：徐文良建筑师事务所；机电：筑友工程顾问有限公司；空调：林一声建筑师事务所

施工方
竟成工程有限公司、雨田工程有限公司

摄影师
钟荣光

全人教育村

项目主持人
潘冀

合作建筑师
王秋华

项目成员（建筑 / 景观 / 室内）
许国、江建平、蔡恒升、唐正国、林镱汰、陈伯桢、马楷、徐健铭、钱云儒、赵珮诗、许嘉麟、蔡祖烈、郭义斌、李鸿勋、吴仁惠

业主
中原大学

顾问
结构：弘运工程顾问有限公司；机电：泉成机电工程有限公司；空调：元泰工程顾问有限公司；灯光：大公照明设计顾问有限公司；音响顾问：江维华教授

施工方
双喜营造股份有限公司、重光行工程股份有限公司

摄影师
郑锦铭

新中北学生宿舍：热诚楼

项目主持人
潘冀

项目成员（建筑 / 景观 / 室内）
苏重威、黄种财、曾彦智、陈伯桢、陈良慈、张晏维、凌琳、陈韦纶、黄英婷、黄诗芳、杨家琳、吴方诚、曹国龙、李孝文、许堡朝、李得嘉

业主
中原大学

顾问
结构：群策工程顾问股份有限公司；机电：吴建兴电机空调技师事务所；空调：元泰工程顾问有限公司；灯光：颉乘设计照明工作室

施工方
双喜营造股份有限公司

摄影师
谢伟士、吴启民、曾彦智

马偕医学院校舍

项目主持人
潘冀

项目成员（建筑 / 景观 / 室内）
黄种财、陈绍平、曾彦智、翁源彬、覃亦仙、苏玮、王素娟、黄诗芳、杨家琳、黄介帅、林柏阳、蔡祖烈刘亿松、陈彦璋

业主
马偕医学院

顾问
结构：筑远工程顾问有限公司；机电顾问：吴建兴电机空调技师事务所；空调：元泰工程顾问有限公司

施工方
利晋工程股份有限公司

摄影师
谢伟士、曾彦智

兴隆公共住宅一区

项目主持人
陈洁生

项目成员（建筑 / 景观 / 室内）
潘冀、陈伯桢、张志明、张元琳、陈扬升、凌琳、蔡慈媛、吴柏璋、黄诗芳

监造
喻台生建筑师事务所

业主
台北市政府都市发展局

统包营造
润弘精密工程事业股份有限公司

摄影师
梁建元、陈扬升、张元琳

财政主管部门土地合作开发项目

项目主持人
郑荣裕

项目成员（建筑 / 景观 / 室内 - 公部门）
潘冀、张晓鸣、陈伟民、梁建元、简雪芬、王素娟、吕理俊、谢绵丽、陈盈洁、黄诗芳、钟安媛、杨家琳、黄彦霖、史晓梅、张钧婷、丁家伟、潘天壹、杨天闿、庄智丞、沈庆豊、魏鸿林、魏绍翔、陈政宏、陈彦璋、杨子莹、王嘉珍

业主
财政主管部门、华固建设股份有限公司

顾问
结构：群策工程顾问股份有限公司；机电空调：吴建兴电机空调技师事务所（公部门）、正弦工程顾问有限公司（住商大楼）；灯光：颉乘设计照明工作室；室内：颉合设计（住商大楼）

施工方
品兴营造股份有限公司（公部门）、大陆工程股份有限公司（住商大楼）

摄影师
小雄梁彦、杨岳航、梁建元

伯大尼少儿家园

项目主持人
潘冀

项目成员（建筑 / 景观 / 室内）
郑荣裕、陈绍平、翁源彬、黄诗芳、赖奕学、张博隆、陈彦璋

业主
基督教台北市私立伯大尼儿少家园

顾问
结构：群策工程顾问股份有限公司；机电空调：吴建兴电机空调技师事务所

施工方
三星营造股份有限公司

校园福音团契大楼

项目主持人
潘冀

项目成员（建筑 / 景观 / 室内）
杨天闿、马法耶、魏名妤、黄诗芳

业主
台北市基督教校园福音团契

顾问
结构：筑远工程顾问有限公司；机电空调：吴建兴电机空调技师事务所

施工方
利晋工程股份有限公司

台大医学院先进医疗基地

项目主持人
苏重威

项目成员（建筑 / 景观 / 室内）
癌症医学中心
潘冀、张晓鸣、陈伟民、王文棋、简雪芬、梁建元、陈盈洁、谢绵丽、黄诗芳、吴淑靖、曾彦智、谢宗芳、毛轩扬、刘彦宏、彭琦瑄、陈奕豪、毛语薇、陈连盛、吴方诚、王秋斌、蔡祖烈、严丹璞

辐射科学暨质子治疗中心
陈伟民、黄彦霖、史晓梅、颜伯菘、唐伟哲、黄诗芳、戴慧如、陈南宏、严丹璞

业主
永龄健康基金会

医院前期规划暨设计
癌症医学中心
美国 HKS 建筑设计公司

顾问
癌症医学中心
结构：群策工程顾问股份有限公司；机电：汉达工程顾问有限公司；空调：元泰工程顾问有限公司；景观：环艺工程顾问有限公司；灯光：颉乘设计照明工作室；室内：颉合设计；幅防：费曼科技股份有限公司

辐射科学暨质子治疗中心
结构：群策工程顾问股份有限公司；机电：明智工程顾问有限公司；空调：元泰工程顾问有限公司；特殊医疗设备：Bard, Rao + Athanas Consulting Engineers, P.C.；BIM：卫武信息股份有限公司；滞洪：元丰工程顾问有限公司

施工方
癌症医学中心
达欣工程股份有限公司

辐射科学暨质子治疗中心
日商清水营造工程股份有限公司台湾分公司

永续

张杨乾 x 苏重威：建筑师与业主的正面协作关系

摄影师
王维康、台达电子文教基金会、刘俊杰、刘冠良、JJP

建筑接地气：台达电子合作系列

永续之环，大溪小学风雨球场

项目主持人
潘冀

项目成员
苏重威、曾彦智、陈伯桢、张晏维、刘冠良

业主
台达电子文教基金会

顾问
结构：筑远工程顾问有限公司

施工方
利晋工程股份有限公司、新伦营造有限公司（球场）、志坚工业有限公司（球场）

摄影师
台达电子文教基金会、曾彦智、刘冠良

台达台南二期厂房

项目主持人
苏重威

项目成员（建筑）
潘冀、洪亮辰、许国、陈武志、蔡佳宏

业主
台达电子工业股份有限公司

顾问
结构：群策工程顾问股份有限公司；机电：高齐能源科技股份有限公司 - 王学训；景观：原庄景观工程有限公司

施工方
利晋工程股份有限公司

摄影师
刘俊杰、台达电子工业股份有限公司、洪亮辰

台达美洲总部

项目主持人
潘冀

项目成员（建筑 / 景观）
许国、李大威、黄诗芳、叶宗舜、吴柏璋、陈扬升

业主
Delta Electronics (Americas) Ltd.

签证建筑师
Korth Sunseri Hagey Architects

顾问
结构：Nishkian Menninger;
机电：Timmons Design Engineers;
景观：SWA Group

施工方
Vance Brown Builders

摄影师
Rick Gardner, 李大威

台中一中校史馆修复及再利用

项目主持人
苏重威

项目成员（建筑 / 景观）
潘冀、陈伯桢、康兆言、陈昭邑、魏嘉诚、刘为杰

业主
台达电子文教基金会

顾问
史迹修复：一元创合设计顾问有限公司 - 林雅萍建筑师；结构：群策工程顾问股份有限公司；机电：上造工程顾问有限公司；灯光：颉乘设计照明工作室

施工方
泰南营造有限公司

摄影师
刘俊杰、台中一中

群裕设计咨询上海办公室

项目主持人
潘冀

项目成员（建筑 / 景观 / 室内）
吴昱廷、郑荣裕、黄智隆、郑仁杰、蔡恒升、吴俊龙、李立凯、李昱霖

顾问
结构：群策工程顾问股份有限公司、上海隆衡装饰有限公司；机电：上海津日机电设备安装有限公司

施工方
上海隆衡装饰有限公司

摄影师
张全箴、莫尚勤、吕恒中、JJP

台积电园区

项目主持人
十二厂、十五厂：潘冀
十四厂：郑荣裕

项目成员（建筑 / 景观）
十二厂总部大楼
陈伟民、周大同、曾文忠、谢宗芳、黄世铭、严纬勇、宋契德、林淑惠、蔡志伟、侯冠如、黄永彬、赖敏政、沈庆丰、谢寿明、栗正暐

十二厂四期
蔡志伟、陈伯桢、洪亮辰、沈建宏、吴顺郎、林崑炤、许舒乔、侯冠如、施钧凯、黄诗芳、黄永彬、苏怡樵、洪宏达、史伟富、游顺益、刘亿松、苏政宏、侯洸伍、徐国浩、陈孟杰、潘永庆

十四厂五期
潘冀、严纬勇、蔡志伟、曾彦智、侯冠如、郑顺益、毛以励、吴淑精、李宗荫、吕理峻、张懿芬、梁向宇、游明宪、康兆言、朱育萱、黄诗芳、李昱霖、沈庆丰、陈聪贤、黄松柏、蔡佳宏、吕国维、颜佑展、许智程、杨家铭、徐培乔、许堡朝、魏绍翔、王腾弘、李家豪、郭孟旻

十五厂一期
陈伟民、李翩、曾彦智、吴轩铭、谢宗芳、吴晓云、刘峻志、王文棋、黄彦霖、杨琇茹、廖志坚、谢绵丽、严纬勇、黄诗芳、李宗荫、吕理俊、吴姿桦、梁向宇、陈扬升、彭琪瑄、沈建宏、刘彦宏、林柏阳、梁世伟、萧琪、廖文瑜、吕昂，蔡政颖、许堡朝、洪宏达、廖哲逸、许智程、刘俊龙、刘为杰、黄松柏、颜佑展、郭孟旻、林鼎翰、魏绍翔、杨家铭、黄驯明、李世民、黄广祺

业主
台湾积体电路制造股份有限公司

顾问

十二厂总部大楼

总顾问：IDC 美商爱迪西股份有限公司；结构：祥谷工程顾问公司；机电：安群工程顾问有限公司；景观：环艺工程顾问有限公司；灯光：大公照明设计顾问有限公司；水保：中基土壤技术顾问有限公司；室内设计：元崇设计工程公司

十二厂四期

结构：群策工程顾问股份有限公司；水保：中基土壤技术顾问有限公司

十四厂五期

结构：群策工程顾问股份有限公司；机电：益鼎策工程顾问公司（办公栋）、汉唐集成股份有限公司（厂房）；景观：老圃造园工程股份有限公司；大地：中基土壤技术顾问有限公司；室内设计：元崇设计工程公司（办公栋）

十五厂一期

结构：群策工程顾问股份有限公司；机电：圣晖工程科技股份公司（办公栋）、汉唐集成股份有限公司（厂房）；景观：老圃造园工程股份有限公司；水保：中基土壤技术顾问有限公司；室内设计：元崇设计工程公司

施工方

十二厂总部大楼

互助营造股份有限公司、力福实业股份有限公司（外墙）、中国钢铁结构股份有限公司（钢构）

十二厂四期

互助营造股份有限公司、大三亿营造股份有限公司（办公区）、达欣工程股份有限公司（连通桥）

十四厂五期

大三亿营造股份有限公司（办公栋）、力福实业股份有限公司（办公栋外墙）、互助营造股份有限公司（厂房）、达欣工程股份有限公司（厂房）、壹东实业股份有限公司（厂房外墙）、中国钢铁结构股份有限公司（钢构）

十五厂一期

达欣工程股份有限公司（办公栋）、美港联合股份有限公司（办公栋外墙）、互助营造股份有限公司（厂房）、壹东实业股份有限公司（厂房外墙）、台湾实拓股份有限公司（厂房外墙）、中国钢铁结构股份有限公司（钢构）

摄影师

陈弘暐、郑锦铭、刘俊杰

从熟龄到乐龄：双连赡养系列与熟龄好宅研究计划

双连教会社会福利园区

项目主持人

一期：潘冀
二、三期：黄种财

项目成员（建筑 / 二三期景观 / 二三期室内）

一期：唐正国、简雪芬、谢宗芳、关惠尤、李志铭、蔡祖烈
二期：唐正国、史晓梅、陈雅祯、诸汉益、蔡祖烈
三期：苏重威、陈伯桢、谢伟士、李翩、吴晓云、洪亮辰、吴轩铭、刘建生、刘亿松、蔡祖烈、曹国龙

业主

台湾基督长老教会双连教会

顾问

一期
结构：祥谷工程顾问公司机电顾问；机电：泰迪工程顾问公司；空调：林伸环控设计有限公司；景观：江堰环境工程顾问公司；室内：汉象设计工程股份公司

二期
结构：群策工程顾问股份有限公司；机电：泉成机电工程顾问公司；空调：林伸环控设计有限公司

三期
结构：群策工程顾问股份有限公司；吴建兴电机空调技师事务所；空调：林伸环控设计有限公司；灯光：徐敏瑜；外墙分析：利道科技工程有限公司（教堂）；音响：江维华教授（教堂）

施工方

一、二期：圣堡营造股份有限公司
三期：晋荣营造股份有限公司、垣登电机股份有限公司

摄影师

曾敏雄、陈弘暐、陈逸文、谢伟士、陈伯桢

双连新庄社会福利中心

项目主持人

黄种财

项目成员（建筑 / 景观）

苏重威、陈伯桢、叶胜钿、黄天佐、刘冠良、杨家琳、许舒涵、刘亿松

业主

新北市双连社会福利慈善事业基金会

顾问

结构：群策工程顾问股份有限公司；机电空调：吴建兴电机空调技师事务所；室内：谢伟士建筑师

施工方

优宝营造有限公司

摄影师

谢伟士

新世代熟龄好宅研究计划

计划主持人

黄种财

计划成员

潘冀、苏重威、黄巾倪、冯如筠、梁廷达、陈扬升、廖志坚、陈伯桢、余启佑、张佩珊、冷雪霜、曾彦智、郑仁杰、杨天闿、黄介帅、胡幼函、陈泽宏

合作伙伴

勤美股份有限公司、璞真建设股份有限公司、勤美璞真文化艺术基金会

建筑师

谢伟士建筑师事务所、林祺锦建筑师事务所

情境导引

应用剧本实验室（余德彰、林文绮、陈学琳）、自由空间教育基金会（唐峰正、王湘淳）

学者专家

龚书章教授、李明达教授、双连赡养中心（蔡芳文、蔡孟炜）、陈坚志

场地提供

真书轩、璞真建设股份有限公司、香格里拉乐园、勤美术馆、空总创新基地

摄影师

曾彦智、陈扬升

共享

有机工作室

摄影师

郑锦铭、吴启民、曾敏雄、谢伟士、王维康、蔡君阳、梁建元、曾彦智、冷雪霜、JJP

手绘

陈平

团队成员

摄影师

汪德范

大事记

摄影师

林淑卿、曾敏雄、郑锦铭、陈弘暐、刘俊杰、谢伟士、陈逸文、黄威博、张晓鸣、曾彦智、JJP

luminocity.cn

光明城

LUMINOCITY

“光明城”是同济大学出版社城市、建筑、设计专业出版品牌，由群岛工作室负责策划及出版，致力以更新的出版理念、更敏锐的视角、更积极的态度，回应今天中国城市、建筑与设计领域的问题。